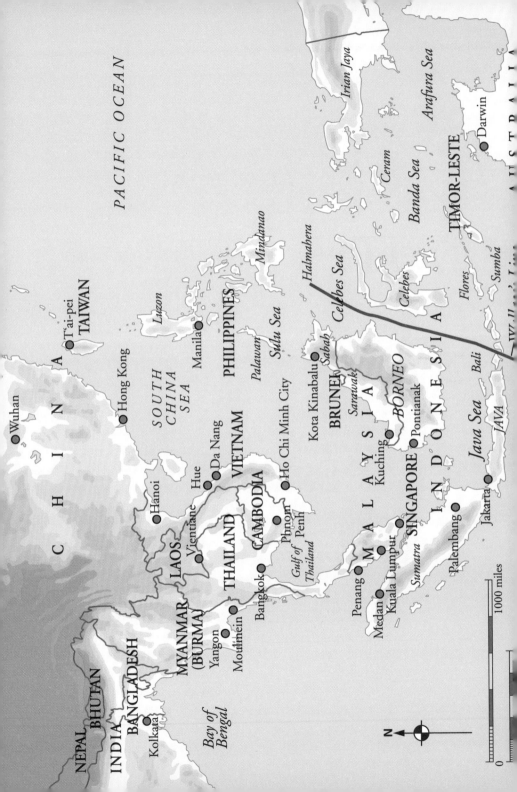

A NATURALIST'S GUIDE TO THE
LIZARDS
OF
SOUTHEAST ASIA

Jordi Janssen & Emerson Sy

First published in the United Kingdom in 2022 by John Beaufoy Publishing Ltd
11 Blenheim Court, 316 Woodstock Road, Oxford OX2 7NS, England
www.johnbeaufoy.com

Copyright © 2022 John Beaufoy Publishing Limited
Copyright in text © 2022 Jordi Janssen & Emerson Sy
Copyright in photographs © see below
Copyright in maps © 2022 John Beaufoy Publishing Limited

Photo Credits
Front cover: *main image Gonocephalus chameleontinus* © Richard Moore
Back cover: *top to bottom: Cyrtodactylus consobrinus* © Steven Wong, *Tytthoscincus hallieri* © Indraneil Das, *Varanus komodoensis* © Jonne Seijdel
Title page: *Gonocephalus doriae* © Wayneson Tay
Contents: *Pseudogekko smaragdinus* © Emerson Y. Sy

Photo Credits
Main descriptions are denoted by a page number followed by t (top), b (bottom).
Bernard Dupont 40; **Dylan van Winkel** 71b, 99b, 145; **Eduard Galoyan** 54b; **Eddy Even** 106b; **Emerson Y. Sy** 9, 19b, 36, 39, 73, 84t, 85b, 90b, 94b, 95b, 96b, 100b, 106t, 127b, 136, 140t, 142t; **Erikson Tabayag** 26b; **Erl Pfian T. Maglangit** 23b, 63t, 100t; **Gregg Yan** 18b; **Gernot Vogel** 14t, 33, 84b, 96b, 97b, 115t, 120b, 125b, 127t, 139, 141b; **Hinrich Kaiser** 99t, 108t, 146; **Indraneil Das** 30, 51, 53, 56, 60b, 70t, 74t, 86, 87b, 102, 112t, 125t, 129t, 131t, 132tb; **Jamie Dichaves** 142b; **Jojo de Peralta** 75b; **Jonne Seijdel** 138; **Jordi Janssen** 14b, 19t, 20, 21b, 34, 45t, 135; **Kenneth Chin** 17b, 35t, 61t, 64tb, 70b, 92t, 98t, 112b, 133; **Kier Mitchel E. Pitogo** 26t, 63b, 107b, 111t, 119t, 130t; **Kirat Kunya** 93t; **Maren Gaulke** 89t; **Nathan Rusli** 28b, 71t, 94t, 116t, 123b, 143; **Nick Baker** 13, 22, 24tb, 41t, 74t, 91t, 96t, 104, 105b, 107t, 108b, 114t, 118t, 121t, 121b, 122t, 122b, 141t, 144t; **Parinya Pawangkhanant** 29, 35b, 46t, 48t, 50tb, 57b, 58t, 59t, 76b, 77t, 81b, 82t, 87b, 103tb, 109, 110tb, 113t, 115b, 118t, 134, 144b; **Paul Freed** 38tb, 42tb, 49, 52, 54t, 55t, 60t, 65t, 67b, 79tb, 88b, 92b, 98t, 114b, 120t, 126t, 137t, 140b; **Peter Brakels** 44t, 69t, 91b, 117t, 117b, 128t; **Peter van Issem** 59b, 65b, 67t, 78; **Richard Moore** 32; **Robin Backhouse** 23t, 43b; **Roderick Parcon** 129b; **Sandra Jye** 58b, 93t, 126b; **Sankar Ananthanarayanan** 37; **Steven Wong** 8b, 17b, 25tb, 27tb, 31tb, 45b, 47b, 66t, 74b, 75t, 85t, 88t, 89b, 95t, 97t, 101, 105t, 111b, 124t, 128t; **Thomas Ziegler** 16b, 62t, 66t, 76t, 80t, 131b; **Thor Håkonson** 55b, 61b, 62b, 69b, 72t, 124b; **Ton Smits** 15, 16t, 17t, 18t, 21t, 28t, 41b, 43t, 46b, 47t, 48b, 68tb, 72b, 77b, 80b, 81t, 82b, 83tb, 90t, 119b, 123t, 130b, 137b; **Virtito Natural Jr.** 116b; **Veronica Prudente** 113b; **Wayneson Tay** 44b.

All rights reserved. No part of this publication may be reproduced, stored in a retrieval system or transmitted in any form or by any means, electronic, mechanical, photocopying, recording or otherwise, without the prior written permission of the publishers.

Great care has been taken to maintain the accuracy of the information contained in this work. However, neither the publishers nor the authors can be held responsible for any consequences arising from the use of the information contained therein.

ISBN 978-1-912081-58-5

Edited and indexed by Krystyna Mayer
Designed by Gulmohur
Project Management by Rosemary Wilkinson
Printed and bound in Malaysia by Times Offset (M) Sdn. Bhd.

·Contents·

Introduction 4

Southeast Asia 4

Lizard Conservation 5

Lizard Identification 7

Searching for Lizards 9

About This Book 10

Glossary 11

Species Descriptions 13

Checklist of the Lizards of Southeast Asia 147

Further Information 173

Index 174

INTRODUCTION

From the Common Water Monitors that roam the waterways, to the geckos hunting and calling around street lights at night, lizards can be some of the most frequently encountered animals in Southeast Asia. As of June 2021, a total of 7,059 lizard species have been recognized worldwide, of which 1,122 (15.8 per cent) have been recorded in Southeast Asia. This makes Southeast Asia one of the most biodiverse areas when it comes to lizard species. In addition, it is home to the largest lizard species in the world, the Komodo Dragon *Varanus komodoensis*. Lizards are featured in Southeast Asia's folklore, mythology, and trade for medicine, meat, skins and live animals. They have historically been regarded as mysterious, and have long been associated with special abilities. A lizard may bring good fortune or it can be the harbinger of bad luck, for instance due to a reputation for spreading diseases (such as skin disorders). Lizards are often used in traditional medicine, sometimes even to cure the disease that the particular species is accused of causing.

This guide covers 225 species, almost 20 per cent of the lizards in Southeast Asia (here consisting of Brunei Darussalam, Cambodia, Indonesia, Laos, Malaysia, Myanmar, the Philippines, Singapore, Thailand, Timor-Leste and Vietnam). Although many of the lizards considered here are endemic to this region, many also occur far and wide outside it. Numerous areas in Southeast Asia are relatively unexplored and new lizard species are still described each year.

The aim of this book is to provide a tool for rapid identification of lizard species. By using the descriptive text and photographs showing colour and morphology, they can be identified by anyone interested in them, regardless of their experience with lizard identification. Scale counts are provided for some species, but in general, when it comes to species for which scale counts are needed for positive identification, further reading is suggested. Some of the species featured have not been illustrated in publications before.

SOUTHEAST ASIA

In this book, the area defined as Southeast Asia comprises two geographic regions, consisting of Mainland Southeast Asia (Cambodia, Laos, Myanmar, Peninsular Malaysia, Thailand and Vietnam) and Maritime Southeast Asia (Brunei Darussalam, east Malaysia [Sabah and Sarawak], Indonesia, Timor-Leste, the Philippines and Singapore). Geographically, the Andaman and Nicobar Islands are also considered to be part of Maritime Southeast Asia, but they are omitted here.

Southeast Asia stretches for roughly 4,500,000km^2, making up almost 3 per cent of the total land mass on Earth. Due to its situation on the intersection of several geological plates, the area has a high occurrence of earthquakes and volcanic eruptions.

CLIMATE

The majority of Southeast Asia has a tropical hot and humid climate with plenty of rainfall. Exceptions to this are northern Vietnam and parts of Laos and Myanmar, which have a more subtropical/temperate climate with dry winters and hot summers. The climate in central Myanmar can even be considered as an arid steppe climate. Seasonality is mainly based on rainfall, with a clear wet and dry season caused by shifts in winds or monsoons.

LIZARD HABITATS

The lizards of Southeast Asia use a wide range of habitats, including sandy beaches and rocky shores, mangrove forests, coastal vegetation, peat swamp forests, (karst) limestone, lowland and hill dipterocarp forests, montane forests, and man-made habitats (like agriculture and other forms of urban development). Due to the large range of available habitats and microhabitats, and the great diversity of potential prey, numerous niches are available to the lizards of Southeast Asia. Many are highly specialized to their local environment. For instance, the karst habitats of Myanmar harbour an enormous diversity of highly endemic and specialized lizards, with often a different species on each mountain. This high diversity is often caused by the fact that numerous microhabitats are available, and that the hills are frequently isolated from each other.

Several lizard species have adapted to living in human-dominated environments and can often be seen running in gardens or on the walls of houses at night. Whereas oil-palm plantations generally support fewer species than forests, Common Water Monitors *Varanus salvator* and Common Sun Skinks *Eutropis multifasciata* seem to thrive in this habitat. Monocultures like tea plantations are often avoided by lizards, but species like Common Sun Skinks do well in this type of habitat. Rubber plantations are great habitats for lizards that require open spaces, like flying dragons (*Draco* species) and butterfly lizards (*Leiolepis* species).

LIZARD CONSERVATION

Of all lizard species in Southeast Asia, 61 per cent are currently evaluated (see Table 1) on the IUCN Red List of Threatened Species (version 2021.1), of which the largest percentage is considered Least Concern (35 per cent). A total of 95 species is considered to be threatened, falling within the categories of Critically Endangered, Endangered or Vulnerable. For roughly 166 species, not enough data is available on distribution and abundance to make an assessment and they are thus classified as Data Deficient.

Southeast Asia is among the most biodiverse areas in the world, with at least four of the 25 globally important hotspots lying within its region. An estimated 15 per cent of the world's tropical forests can be found in Southeast Asia. Indonesia (between Borneo and Sulawesi, and between Bali and Lombok) encompasses the Wallace Line, and the faunal boundary line separates the biogeographic areas of Asia and Australia. West of

Lizard Conservation

	Species	CR	EN	VU	NT	LC	DD	NE
Agamidae	174	2	3	9	4	60	27	69
Anguidae	6					4	1	1
Dibamidae	21		1		1	5	8	6
Eublepharidae	6	1	1	1		1		2
Gekkonidae	472	11	17	35	14	166	57	172
Lanthanotidae	1							1
Pygopodidae	2					2		
Lacertidae	5					3	1	1
Scincidae	394	2	3	4	13	141	68	163
Shinisauridae	1		1					
Varanidae	40		2	2	1	9	4	22

TABLE 1 Conservation status of lizard species occurring in Southeast Asia (IUCN Red List of Threatened Species, Version 2021.1.) Only CR, EN and VU are considered threat categories; the rest are non-threatened categories. **CR** Critically Endangered; **EN** Endangered; **VU** Vulnerable; **NT** Near Threatened; **LC** Least Concern; **DD** Data Deficient; **NE** Not Evaluated.

the line, species can be found that are primarily related to Asiatic species, while east of it, a mixture of species of Asian and Australian origin is found. The region is also among the world's top deforestation hotspots. Habitat loss in Southeast Asia is among the highest and most severe when considering biodiversity loss.

Between 2005 and 2015, Southeast Asia lost approximately 80 million hectares of forest, of which Indonesia accounted for roughly 62 per cent, followed by Malaysia (16.6 per cent), Myanmar (5.3 per cent) and Cambodia (5 per cent). Loss of habitat can have big impacts on the lizard species in this area. While certain species do very well in or near human settlements, or in – for instance – oil-palm plantations, not all species can survive. Loss of habitat is one of the main threats to the survival of many lizard species, in particular endemic ones or those confined to a very narrow range and/or having specific habitat requirements.

Besides the threat of habitat loss, lizards face several other threats. They play an important cultural and economical role for many in Southeast Asia, and are harvested in large volumes, both legally and illegally, for human consumption, skins, and use as pets or in traditional medicine. They are often used for multiple purposes.

Lizards have been widely used to treat a large variety of ailments such as coughs, asthma, diabetes, cancer, hyperglycaemia, skin ailments, bruises, sprains, arthrosis and rheumatism. The medicinal use of many species is still undocumented, and trade for this purpose largely stays within the boundaries of a country. One exception is the trade in

Lizard Conservation

Tokay Geckos *Gekko gecko*, which are harvested in the millions each year to be dried and used in traditional medicine. Concerns about the impact of this trade on the population resulted in a listing on the Convention on International Trade in Endangered Species of Wild Fauna and Flora (CITES), requiring trade in the species to be regulated.

The global trade in reptile skins represents the largest volume for all uses of reptile species. One of the largest lizards in Southeast Asia, the Common Water Monitor, a widespread species, plays a significant role in this trade. This species is considered among the top five most frequently traded CITES-listed reptile species for the trade in skins. Between 2000 and 2017, about 13–15 million skins of the species were traded internationally. Singapore, Indonesia and Malaysia are all listed among the largest exporters and importers of reptile skins in the world. The trade in reptile skins is thus an important potential threat for the species involved.

The human consumption of lizards is significantly related to cultural beliefs and perceived medicinal value. Trade for these purposes overlaps. The perceived medicinal value of lizards can be traced back to Java during the Pleistocene era. Today, live lizards and their parts may be purchased in markets throughout Southeast Asia, where they are boiled, consumed in the form of satay or barbecued. This often concerns larger lizards like monitor lizards, but also involves smaller skinks and geckos.

In addition, many lizard species are captured alive for the global trade in pets. Newly described or rare species are particularly desired by collectors. The international trade in live pets impacts more species than any other use, making it a serious threat. This is particularly the case when it is combined with additional threats that place further stressors on a species. The combined impacts of habitat loss and trade are probably severely underestimated.

The Southeast Asian region still harbours many areas that have been poorly explored when it comes to the lizard diversity. New species are still described at a rapid rate. In Myanmar, 24 new lizard species have been described in recent years, with a further 14 still waiting to be described. In particular, isolated pockets of karst limestone habitat appear to harbour many newly discovered species, often confined to a particular pocket or cave. The very high biodiversity in some of these ecosystems suggests that many more species still remain undiscovered, and that not enough information is available to assess their conservation status.

Lizard Identification

Morphological details like colouration, pattern and certain scales allow many lizard species to be identified. Features that are particularly useful for their determination are types of scale (for example keeled or smooth), and the presence of nuchal/dorsal crests, spines, patagia and tympanum. The size of a lizard is primarily determined by measuring the snout–vent length (SVL) rather than the total length (TL) – because many lizards have the ability to autotomize their tails, subsequently regenerating them, measuring the total length is of little use for identification.

Lizard Identification

However, the majority of this book's users will use colour and form to identify species. As many lizards behave skittishly around people, binoculars can be useful in observing detailed morphological features from far away enough to avoid disturbing the lizards.

Not all species can be easily identified based on colour, form or other morphological features, and the counting of scales is often required to confirm identity. More detailed technical works should be consulted in the case of species to which this applies.

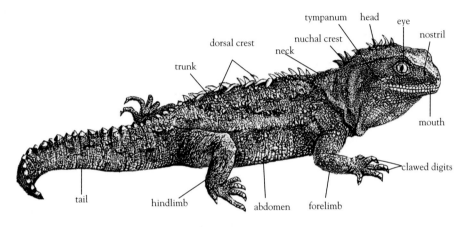

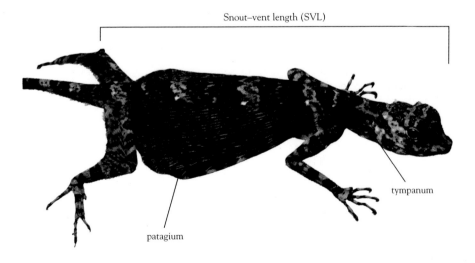

SEARCHING FOR LIZARDS

Lizards can be very easy to locate in Southeast Asia. Several gecko species (primarily *Hemidactylus*) can often be found on walls of buildings in the evening, hunting for insects around street lights. The distinctive call of the Tokay Gecko can be heard in many places, in particular around human settlements. Water monitors can be a common sight in certain parks or around any waterbody. Locating lizards depends a lot on the type that is being searched for. There are lizards that are primarily nocturnal (many of the geckos), those that are diurnal and frequently bask (such as *Leiolepis* and *Calotes* species), and others that are diurnal but more easily seen during the night (like *Gonocephalus* species).

Lizards are often noticed because they run away when they see a human and are then either seen or heard. Often, if lizards are disturbed while basking, they will return after a bit of time, especially if you sit silently and do not move. However, more cryptic species can be hard to find, and are either located by luck, or for instance by turning stones or logs. At night, lizards may be seen sleeping on branches while spotlighting (using a flashlight, although doing this so that the light focuses directly on any animal's eyes should be avoided). In addition, certain genera of Southeast Asian lizards, such as *Draco* species and some geckos, are able to parachute themselves or glide through the air to safety.

Note that fallen logs, stones and other potential hideouts might lose their function for lizards and other animals if not placed back carefully in the same positions that they were in before being moved. In addition, venomous snakes may be present in these hideouts, occurring in the same habitats that many lizard species frequent. While searching for lizards, never just put a hand inside a crack or hole, or under an item where snakes may be sheltering.

LIZARD DEFENCE MECHANISMS

Many lizard species have evolved threat postures or defence mechanisms to deter or escape from potential predators, including humans. The blue-tongued skinks (*Tiliqua* species), of which two species occur in Indonesia, have broad blue tongues that they readily protrude from their wide mouths when they feel threatened. Many arboreal lizards drop from their perches when disturbed, either gliding or parachuting their way to safety. Other lizards drop into nearby streams and try to escape by staying underwater.

Caudal autotomy, or the voluntary shedding of the tail in response to a threat, is a common defence mechanism in lizards. It is particularly common in smaller species, which often have brightly coloured tails. When threatened or

Tiliqua gigas

About This Book

when the tail is grabbed, the lizard will voluntarily drop its tail, after which the tail will wriggle furiously, distracting the potential predator. The tail may grow back, but will look different from the original tail. Note that the tail often contains important energy reserves for lizards, and that losing it can have serious consequences for the health of a lizard.

However, not all species can shed their tails, and some use them as weapons – monitor lizards, for example, will try to hit a threat with the tail. Then there is another gecko (*Gehyra mutilata*) in Southeast Asia that uses dermal autotomy as a defence mechanism. In this defence mechanism, the skin of the lizard sits loosely around the body, and tears and comes off when grabbed by a predator or human. While the lizard may get away, leaving the predator with torn skin in its mouth or hand, this will leave the lizard vulnerable to infection and water loss while the skin regenerates. A lizard should therefore never be caught by grabbing it by the tail or body when you are not familiar with a species.

Note that some lizard species are protected in their range states, and that disturbing, grabbing or killing a lizard can be a criminal offence without the required permits.

About This Book

This book deals with a representative number of lizard species that can be found in Southeast Asia, as defined above (p. 4). Some are very common, others very rare and difficult to find. The aim of the book is to provide rapid identification for anyone interested in lizards, regardless of their experience with lizard identification. Roughly 20 per cent of all lizard species (1,122 species, June 2021) that occur in Southeast Asia are described. The distribution information allows anyone to identify lizard species, or gain enough information to continue a targeted search elsewhere in the literature.

For each species, the common English name (if available) is provided, as well as the scientific name and the maximum snout–vent length. Species descriptions focus on colours and morphological features that aid identification. Scalation details are sparsely given, primarily where they are required to positively identify a species. Also provided are distribution ranges, including doubtful records, and notes on habitat, elevation range and habits of the species. Data on diet and reproduction is sparsely given. The conservation status of each species is provided in the checklist of species (p. 147), with a summary being given in the introduction.

■ Glossary ■

Glossary

anterior Position towards front.
arboreal Living above the ground or in trees, vegetation or rocks.
asl Above sea level.
autotomy Ability of a species to break off a body part, usually the tail, when threatened or stressed.
axilla Armpit.
carnivorous Feeding on other animals.
caudal Refers to the tail or towards tail region.
cloaca Chamber into which intestinal, urinary and reproductive ducts discharge contents.
clutch Set of eggs laid by female at one time.
crepuscular Active during dusk and dawn.
crest Ridge on head, neck (nape) or back consisting of modified scales or skin.
cryptic Camouflaged or hidden.
dewlap Skin-flap under throat of certain lizards, also called gular pouch.
digit Finger or toe.
dipterocarp forest Forest with trees primarily of Dipterocarpaceae family.
distal Further away from point of origin.
diurnal Active during the day.
dorsal Towards upper surface of body.
dorsum Upper side of an animal.
endemic Refers to species occurring only in a restricted geographical region or country.
femoral pore Opening on underside of thigh.
fossorial Living underground or burrowing.
frontal Large median unpaired scale between eyes.
frugivorous Feeding on fruits.
gular Pertaining to or located on throat.
gular fold Transverse fold of skin on throat.
herbivorous Feeding on plants.
introduced Refers to species accidentally or intentionally brought into an area where it does not naturally occur; also known as non-native or exotic.
keeled Refers to scales that each show ridge down centre; sometimes multiple ridges.
labial Pertaining to lip.
lamella (pl. lamellae) Transverse pad underneath digit of lizard.
lateral Pertaining to side of body.
nape Back of neck.
nasal Scale on side of head containing nostril opening.
native Refers to species occurring naturally in a geographical region; also known as indigenous.
nocturnal Active during the night.
nuchal Along back of neck.
ocellus (pl. ocelli) Rounded, eye-like spot.

■ Glossary ■

omnivorous Feeding on both animals and plants.
osteoderm Bony deposit in scale or dermal layer.
oviparous Laying eggs.
ovoviviparous Reproducing by retaining eggs in female's body and giving birth to live young.
parthenogenetic Referring to reproduction without fertilization.
patagium (pl. patagia) Extended skin between front and back limbs, serving as wing in gliding animals.
posterior Position towards the back.
preanal Situated in front of anal opening or cloacal region.
precloacal pore Structure situated in front of cloaca.
preocular Anterior to eye.
primary forest Undisturbed forest.
proximal Close to point of origin.
ritual combat Refers to an intraspecies fight involving two or more individuals to determine the dominant one.
rostral Scale at tip of snout.
rostrum Most anterior part of head snout.
saxicolous Inhabiting rocky areas.
scalation Pattern of scales on body.
scales Small, thin, horny or bony plates covering skin.
secondary forest Previously disturbed forest with new natural growth.
species complex Closely related group of species.
subcaudal Underside or ventral surface of tail.
subfossorial Burrowing.
supranasal Scale above nasal.
SVL (snout–vent length) Measurement from snout-tip to anal opening in straight line.
terrestrial Living on surface of the ground.
TL (total length) Measurement from snout-tip to tail-tip in straight line.
tubercle Nodule on skin's surface.
tympanum Eardrum.
ventral Towards abdomen or underside of an animal.
vertebrals Mid-dorsal row of scales.

▪ AGAMIDS ▪

AGAMIDAE (AGAMIDS)
This group of diurnal and mostly arboreal lizards currently comprises 174 species in Southeast Asia. Agamids are often seen sleeping on branches or on tree trunks at night. They range from the smaller *Draco* species to the large *Hydrosaurus* species, and are related to iguanas. Many Southeast Asian species are often found near waterbodies, but occur in a large range of habitats. Their hindlegs are usually longer than their front legs, and their scales lack osteoderms. Many species have large nuchal or dorsal crests, colourful gular pouches, patagia or rostral appendages. Most feed mainly on arthropods, although larger species are also partially herbivorous.

Peninsular Horned Tree Lizard
▪ *Acanthosaura armata* SVL 140mm

DESCRIPTION Forest-obligate lizard with short, angular head. Pair of recurved spines above eyes. Nape and dorsum with crest. Colour highly variable, from green to brown. Body, tail-base and limbs have scattered, lighter coloured oval markings. DISTRIBUTION Occurs in Indonesia, Malaysia, Myanmar and Thailand. HABITS AND HABITAT Arboreal species usually seen clinging in face-up position on tree trunks about 2–3m above the ground. NOTE Historical records from Singapore are uncertain. Records from Myanmar are questioned, potential misidentification with *A. crucigera*.

AGAMIDS

Green Pricklenape
- *Acanthosaura capra* SVL 137.9mm

DESCRIPTION Green- or olive-coloured dorsum with black spots or yellow spots encircled with black. Throat black with yellow gular pouch in male, greenish in female. Easily distinguished from other *Acanthosaura* species by having only one pair of postorbital spines; lacks nuchal spines. **DISTRIBUTION** Found in hills of eastern Cambodia, and Khanh Hoa and Lam Dong Provinces in southern Vietnam. Records from Laos need validation. **HABITS AND HABITAT** Like all *Acanthosaura* species, diurnal and arboreal, spending most of its time on trees. Clutches of eggs relatively large, with up to 20 eggs. Occurs in lowland and mid-hill evergreen forests to 500m asl.

Cardamom Mountain Horned Agamid
- *Acanthosaura cardamomensis* SVL Female: 149mm; male 133mm

DESCRIPTION Large, with robust body, triangular in shape. Dorsal surface of head, body and limbs greenish-yellow or cream. Top of head of lighter than body, with black eye-patch extending from nostrils through orbit to nape. Dorsum covered in dark marbled/mottled pattern that fades towards tail, but can be very faded. Dewlap in both sexes yellowish or dark. Single cylindrical spine above eye, with long spines in nuchal and dorsal crest. **DISTRIBUTION** Occurs in Cardamom Mountains, from Bokor National Park, Cambodia, to Khao Yai National Park, Nakhon Ratchasima Province, Thailand north-west of Cardamom Mountain range.

HABITS AND HABITAT Seen in forested areas at 400–1,400m asl, both at night and by day. At night sleeps on sapling trees above forest floor, or on vines near stream edges. By day found on leaf litter, on sapling trees, under pieces of bark in leaf litter or on small logs near the ground. Juveniles appear to occur closer to the ground than adults. When provoked, hides or climbs up trees. Oviparous, laying 18–23 eggs and possibly breeding year round.

◾ AGAMIDS ◾

Boulenger's Pricklenape ◾ *Acanthosaura crucigera* SVL 140mm

DESCRIPTION Greenish-yellowish dorsum, and characterized by large nuchal crest. Dorsum covered in dark marbled/mottled pattern. Sides of head dark brown to black, with yellow to white lips and throat. Dark brown to black nape-patch. Tail may be banded. Differentiated from the Cardamom Mountain Horned Agamid (opposite) by significantly smaller spines in nuchal crest, significantly fewer infralabials (10 v 10–14), and fewer subralabials (10–12 v 11–15). **DISTRIBUTION** Found in west Thailand, Myanmar and Peninsular Malaysia. Once the most widespread *Acanthosaura* species in Southeast Asia, but recent taxonomic research found that populations in Cambodia and eastern Thailand belong to the Cardamom Mountain Horned Agamid. Populations in southern Thailand and Peninsular Malaysia believed to belong to five species, of which only two have been described to date (*A. bintangensis* and *A. titiwangsaensis*). **HABITS AND HABITAT** Lays clutches of up to 18 eggs. Occupies lowland evergreen, deciduous and montane forests, where associated with trees. Occurs at 200–1,800m asl.

■ AGAMIDS ■

Brown Pricklenape ■ *Acanthosaura lepidogaster* SVL 101mm

DESCRIPTION One of the more colourful *Acanthosaura* species in Southeast Asia. Dorsum can be brown but can change to bright green; often has black marbled-like markings. Forehead black. Nape has dark diamond shape, usually encircled by white stripe. Lips can be yellow-greenish with orange-reddish, but may also be white – colours usually extend to throat and chest. Nuchal crest has six conical scales. **DISTRIBUTION** Found in Cambodia, Laos, Myanmar, northern Thailand, and north and central Vietnam. Appears to include several undescribed species. **HABITS AND HABITAT** Diurnal and arboreal; also found on the ground. Occurs in lowland and submontane forests at 700–1,400m asl.

Natalia's Spiny Lizard ■ *Acanthosaura nataliae* SVL 158mm

DESCRIPTION Male has yellowish-brown dorsum that can change to shades of red, brown and yellow. Head, nuchal and dorsal crests, and limbs red, ranging from very bright to barely visible. Gular pouch red with 3–5 black bands. Head covered with black mask and postocular stripe. Tail has 9–10 wide dark bands. Female has emerald-green dorsum and similar markings to male. **DISTRIBUTION** Found in southern Laos (Saravan and Xekong Provinces) and central Vietnam. **HABITS AND HABITAT** Diurnal and arboreal. Adults often seen on tree trunks; juveniles found more on bushes and closer to the ground. Lays clutches of up to 16 eggs. Occurs in wet evergreen forests at 350–1,400m asl.

■ Agamids ■

Phuket Horned Tree Agamid
■ *Acanthosaura phuketensis* SVL 123.5mm

DESCRIPTION Medium-sized agamid with single long spine above eye. Second large spine between tympanum and nuchal crest. Nuchal and dorsal crest are strongly developed. Black eye-patch that extends all the way to nuchal crest. Dorsal crest decreases in size towards tail. May have marbled pattern on laterals that decreases towards tail into striped pattern. **DISTRIBUTION** Found on Phuket Island, Thailand. So far only recorded near Khao Phra Thaeo non-hunting area, but probably also occurs in other areas on the island. **HABITS AND HABITAT** Diurnal and arboreal. Can be locally abundant and found in mature secondary forests. At night can be seen asleep on large trees. While primarily arboreal, it is also regularly found on the ground or hiding under rocks near streams.

Earless Agamid ■ *Aphaniotis fusca* SVL 70mm

DESCRIPTION Characterized by bright blue inside of mouth and lack of external ear opening (tympanum). In some populations, iris of male can be a bright blue; iris of female and juveniles brown. Base colour light to dark brown, and may have mottled brown and orange markings. **DISTRIBUTION** Found in southern Thailand, Peninsular Malaysia and Singapore. Also Borneo (Sabah) and possibly Indonesia (including Kalimantan). **HABITS AND HABITAT** Tree and bush dwelling, arboreal and diurnal. Mostly occurs in primary lowland forests and hilly regions. Mainly found near running water on sides of trees and vines. Locally can occur in very high densities, but seems to compete with *Draco* species.

◾ AGAMIDS ◾

Burmese Green Crested Lizard
◾ *Bronchocela burmana* SVL 93mm

DESCRIPTION Slender-looking lizard typical of *Bronchocela* genus. Short nuchal crest of 6–9 scales. Overall greenish colour that can be lighter or darker. Greenish-white cheek-patch. Trunk has transverse band of small green-white spots. Tail has closely spaced white transverse bands. **DISTRIBUTION** Currently only confirmed for Myanmar (southern Tanintharyi Division). May be present in west Thailand. **HABITS AND HABITAT** Diurnal and arboreal. Often found in lowland forests on outer ends of branches near stream sides.

Green Crested Lizard ◾ *Bronchocela cristatella* SVL 120mm

DESCRIPTION Primarily bright green, but can quickly change to brown or grey when threatened or stressed. Very long tail, about three times length of SVL. Head wedge shaped with moderate-sized dark brown ear openings. Low crest from neck to back area. **DISTRIBUTION** Found in Indonesia, Malaysia, the Philippines and Singapore. **HABITS AND HABITAT** Oviparous; female typically lays two long, spindle-shaped eggs. Arboreal, inhabiting lowland primary forests, but thrives in disturbed areas such as parks and gardens. Often seen on low hedges basking by day.

▪ Agamids ▪

Maned Forest Lizard ▪ *Bronchocela jubata* SVL 140mm

DESCRIPTION Robust body with large head. Base colour pale to bright green with faint striped markings; can have yellow or reddish spots on flanks or tail. Male slightly bigger than female. Nuchal crest well developed; dorsal crest barely present. **DISTRIBUTION** Found in Indonesia (Sumatra, Palau Nias, Kalimantan, Sulawesi, Java, Bali and several smaller islands). In the Philippines found on Mindanao and in Thailand around Prachin Buri. **HABITS AND HABITAT** Diurnal. Oviparous, laying clutches of two eggs. Occurs in both primary and secondary lowland rainforests, but also in heavily disturbed areas. Presence of tall trees seems a requirement, and this species can often be seen high up in a tree.

Marbled Green Crested Lizard
▪ *Bronchocela marmorata* SVL 125mm

DESCRIPTION Slender-looking lizard typical of *Bronchocela* genus. Prominent nuchal crest higher than dorsal crest. Overall greenish in colour, which can be lighter or darker. Gular patch well developed. White labials. Darker marbled pattern over body, which can resemble flecks, spots or vertical bands. Darker orbital ring. **DISTRIBUTION** Occurs in the Philippines (Luzon, Polillo, Mindoro, Catanduanes and Sibuan). **HABITS AND HABITAT** Diurnal and arboreal. Commonly observed on low vegetation during the day in disturbed areas. Occurs from sea level to 800m asl.

■ AGAMIDS ■

Forest Crested Lizard ■ *Calotes emma* SVL 75mm

DESCRIPTION Brownish-olive base colour, with darker bands across back, interrupted by lateral white band on each side. Adult male may be very dark in colour, with dark dewlap, and dark eye-patch interrupted by light brown/reddish lips. Female lighter than male, with brown base colour and black markings on head radiating from eye, and two lateral bands with darker markings. Very long toes on front and rear feet; well-developed dorsal crest with spines increasing in size towards nape. Two single spines above eye and two above tympanum (external ear drum). Very variable and can include various shades of brown, green and grey. **DISTRIBUTION** Found in Cambodia, Laos, Peninsular Malaysia, Myanmar, Thailand and Vietnam. Subspecies *C. e. alticristatus* restricted to north Thailand. **HABITS AND HABITAT** Terrestrial and arboreal. Mainly seen by day, although also seen sleeping on leaves and branches during the night. Oviparous, laying about 10–12 eggs. Inhabits wide range of habitats, including dry deciduous, coastal and moist evergreen forests. Often seen on trees, not too far above the ground.

▪ AGAMIDS ▪

Blue Crested Lizard ▪ *Calotes mystaceus* SVL 140mm

DESCRIPTION Typical member of *Calotes* genus, with large head and swollen cheeks. Mainly greyish-brown to reddish-brown base colour; large reddish-brown blotches on flanks. During breeding season, male's head striking blue to turquoise on throat, radiating to back, and reaching to hindlegs. Lengthways pale strip from nostril towards shoulder and passing beneath eye. Like other *Calotes* species, this one has dorsal crest starting above tympanum towards tail-base. Very similar to the **Vietnamese Blue Crested Lizard** *C. bachae*, which occurs in Vietnam. DISTRIBUTION Found in Cambodia, Thailand, Myanmar and Laos. HABITS AND HABITAT Diurnal and oviparous. Fully arboreal. Can be seen at forest edges, and in parks and gardens. NOTE This species has been split after the taxonomic cut-off date for this edition. *C. mystaceus* is now restricted to coastal southern Myanmar, and the species shown here (*C. goetzi*) is found in Cambodia, Myanmar, Thailand and Laos.

Oriental Garden Lizard
▪ *Calotes versicolor* SVL 100mm

DESCRIPTION Displays incredible variation across its range. Crest in adults extending from nuchal towards tail-base. Base colour usually brown, but can be olive or grey with irregular brown markings. Often has radiating dark markings around eye. Breeding male develops bright red or orange patches around head; often black patch just in front of front legs. Can be distinguished from the Forest Crested Lizard (opposite) by lack of spines above eye. DISTRIBUTION Widely distributed in Southeast Asia. Found in Vietnam, Myanmar, Thailand, Peninsular Malaysia, Cambodia, Laos and Indonesia; introduced to Borneo, Singapore and the Philippines. HABITS AND HABITAT Diurnal. Oviparous, laying up to 33 eggs, depending on size of female. Commonly seen in parks, gardens, agricultural areas and open forests. Juveniles tend to forage and bask on the ground; adults often seen on tree trunks.

AGAMIDS

Blanford's Flying Lizard
■ *Draco blanfordii* SVL 120mm

DESCRIPTION Largest member of the genus. Body light brown to grey with olive-grey mottling. Patagium pale yellow and brownish-red on edge (male), or olive-brown with dark radial bands (female). Dewlap light grey and slightly tapered to broad (swollen) tip. Lappet/wattle black proximally and red distally. Male has low caudal crest. **DISTRIBUTION** Occurs in Malaysia, Myanmar, Thailand and Vietnam. **HABITS AND HABITAT** Diurnal and arboreal. Feeds primarily on ants and opportunistically on other small insects. Occurs from lowland rainforest to 1,200m asl.

AGAMIDS

Horned Flying Lizard
Draco cornutus SVL 88mm

DESCRIPTION Dimorphic flying lizard. Female larger (SVL 88mm) than male (SVL 74mm). Thorn-like scale on top of eye. Dorsal head has round black spot between eyes and at base of skull. Dorsum of male vivid green; female brownish-tan. Patagium has black spots forming bands. Patagium margin may be dark reddish-orange. DISTRIBUTION Found in Brunei Darussalam, Indonesia and Malaysia (Borneo). HABITS AND HABITAT Diurnal and arboreal. Occurs at 500–700m asl.

Chartreuse-spotted Flying Lizard
Draco cyanopterus SVL 95mm

DESCRIPTION Active arboreal agamid with small head and slender body. A pair of enlarged thorn-like scales are present above eyes. Dorsal body is grey, tan or brown. Patagium is supported by six ribs. Dorsal patagium has large yellowish-green patches between ribs (male), or dark reticulation or mottling over peach or orange base colour. Triangular dewlap of males is reddish-brown with yellowish to orange tip. Tail has dark brown bands. DISTRIBUTION Found in the Philippines. HABITS AND HABITAT Diurnal, often observed on tree trunks in open areas such as coconut grooves and forest edges, foraging on ants and other invertebrates.

◾ AGAMIDS ◾

Red-bearded Flying Lizard ◾ *Draco haematopogon* SVL 90mm

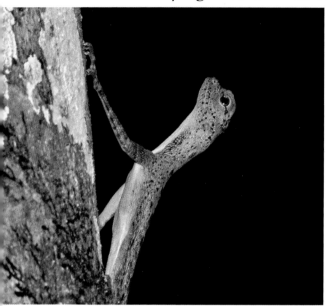

DESCRIPTION Dewlap elongated, triangular, tapering gradually, and covered with small scales; red at base and yellowish-green at tip. Dewlap scales near tip comparatively large. Patagium has five ribs and dark with small light spots. Lappet reddish-orange. Ventral body pink anteriorly and cream posteriorly. **DISTRIBUTION** Found in Indonesia, Malaysia and Thailand. **HABITS AND HABITAT** Diurnal and arboreal. Usually seen on moderate-sized trees and slender saplings, foraging on ants.

Spotted Flying Lizard ◾ *Draco maculatus* SVL 80mm

DESCRIPTION One of the smaller flying lizards in the region, with yellow, orange or pink upper patagium and black spots. Lower patagium yellow with black spots. Dewlap light blue at base and yellow/orange distally. Dorsum light brown to grey. **DISTRIBUTION** Found in Laos, Malaysia, Myanmar and Thailand. **HABITS AND HABITAT** Diurnal and arboreal. Feeds primarily on ants. Female lays 4–5 eggs per clutch.

■ AGAMIDS ■

Giant Gliding Lizard ■ *Draco maximus* SVL 139mm

DESCRIPTION One of the few *Draco* species without sexual dimorphism of patagonium. Dark brown/black base colour with green radial stripes on head, nuchal, dorsum, tail and legs. About three green bands on dorsal. Gular flag dark in colour and can have white markings. Head mainly green with dark brown or black markings. Patagium dark brown or black with lighter lengthways stripes. **DISTRIBUTION** Found in Malaysia, Thailand and Indonesia (Sunda, Sumatra, Natuna and Kalimantan). **HABITS AND HABITAT** Diurnal, sleeping on tree trunks at 1.5–2m off the ground. Oviparous; lays 1–5 eggs. Mainly inhabits forests to 1,000m asl, and can be found around river edges.

Black-bearded Flying Lizard ■ *Draco melanopogon* SVL 90mm

DESCRIPTION Male identified by long, mainly black dewlap. Posterior base of dewlap white to white and orange. Lappet white. Dorsum greyish-green; ventral body cream, with or without round spots. Patagium dark with scattered yellow-orange spots. **DISTRIBUTION** Found in Brunei Darussalam, Indonesia, Malaysia, Singapore and Thailand. **HABITS AND HABITAT** Diurnal and arboreal. Mainly feeds on ants and termites. Female sexually mature at 6.7cm SVL, and lays two eggs, Occurs to 800m asl.

Agamids

Mindanao Gliding Lizard — *Draco mindanensis* SVL 105mm

DESCRIPTION Male has pale brown dorsum with slight greenish cast, and pale or darker brown spots or mottling. Tail and limbs have distinct brown banding and may each enclose pale central band. Neck and gular region may have white spots. Dewlap tangerine-orange. Female creamish on tip of smaller dewlap. Patagium bright red with obscure paler red spots over outer two-thirds, and pale tan striations parallel to body. In female, patagium black with white spots over outer two-thirds, and pale striations on remaining one-third. Ventral surface of patagium in male is red. DISTRIBUTION Found in the Philippines (Dinagat, Leyte, Mindanao and Samar). HABITS AND HABITAT Appears restricted to primary and potentially mature secondary forests. Has been found on large dipterocarp trees in primary forests, often at great height above the ground. Does not appear to be present in coconut groves.

Palawan Gliding Lizard — *Draco palawanensis* SVL 84mm

DESCRIPTION Dorsal pattern brown, black and tan pigments with pale grey base colour, giving mottled appearance. Tail banded with dark brown and pale grey. Male and female have melanic interorbital spot, with only female also having melanic nuchal spot. Female has radiating melanic lines in orbital region. Gular pattern in male brown mottled on pale tan base colour, with one-third of dewlap appearing grey-brown. Dewlap appears orange during display. Dorsal patagial colouration dull orange with several pale longitudinal striations. Distinguished from other *Draco* species by unique patagial colour pattern of dull orange or yellow, with rectangular black spots increasing in size near patagial margin. DISTRIBUTION Only known from island of Palawan in the Philippines. HABITS AND HABITAT Like other *Draco* species, male often seen flicking dewlap and patagia, and producing push-up displays with front legs. Common on trunks of coconut and *Casuarina* trees, and often found in heavily disturbed habitats. Otherwise occurs in dipterocarp rainforests, in edges or canopies.

AGAMIDS

Five-lined Flying Lizard
■ *Draco quinquefasciatus* SVL 110mm

DESCRIPTION Female of this distinctive flying lizard more colourful than male. Body green with small, round dark spots. Patagium yellow proximally and deep orange distally (female) or green (male) with five broad radial bands. Male has elongated yellow dewlap covered with small scales, and tapering to narrow tip. Lappet bright yellow. Patagium has six ribs and transverse bands. DISTRIBUTION Occurs in Brunei Darussalam, Indonesia, Malaysia, Singapore and Thailand. HABITS AND HABITAT Diurnal and arboreal. Female lays up to four eggs per clutch. Usually seen in lowland rainforests on lower tree trunks below 6m above the ground.

Sumatran Gliding Lizards
■ *Draco sumatranus* SVL 90mm

DESCRIPTION Light to dark brown mottled. Patagium black/brown mottled. Black can look like stripes from a distance. Male has triangular gular flag that is mostly yellow with white base. Base can be mottled. Forehead of male often bluish/turquoise. Female has small blue gular flag. DISTRIBUTION Found from southern Thailand to Peninsular and east Malaysia, Singapore and Indonesia (Sumatra, Kalimantan and surrounding islands); possibly in Palawan. HABITS AND HABITAT Diurnal and often seen on trees. Female only comes down to forest floor to lay eggs (2–5 per clutch). Mostly found in disturbed or man-made habitats with plenty of sunlight; rarely in primary or mature forests.

▪ AGAMIDS ▪

Barred Gliding Lizard
▪ *Draco taeniopterus* SVL 83mm

DESCRIPTION Slender-looking lizard with pale grey or green-brown dorsum. Patagium pale grey or green-yellow with five irregular black bands. Throat and dewlap deep crimson, with dewlap dull yellow in male. Ventral area pale greenish or yellowish. **DISTRIBUTION** Found in Cambodia, Peninsular Malaysia (Perlis State), Myanmar (Shan and Mon States), and Thailand. **HABITS AND HABITAT** Diurnal and arboreal. Strongly associated with large trees. Often found at 3–10m above the ground. Inhabits open evergreen forests.

Common Gliding Lizard ▪ *Draco volans* SVL 90mm

DESCRIPTION Looks similar to the Sumatran Gliding Lizard (p. 27). Both sexes have brown base colour with dark flecks on dorsum and nape. Male, similarly to Sumatran, has yellow gular flag, and female a smaller, bluish gular flag. Top side of patagium tan to orange in male and can have dark radial bands. Female patagium lacks dark bands but has irregular dark markings. **DISTRIBUTION** Formerly thought to range throughout Southeast Asia; taxonomic data now only refers to it for the lizards on Java and Bali (Indonesia). **HABITS AND HABITAT** Similarly to Sumatran, often found in forest edges. Favours dry open secondary forests.

◼ AGAMIDS ◼

Abbot's Anglehead Lizard ◼ *Gonocephalus abbotti* SVL 143mm

DESCRIPTION Large, robust agama with lower dorsal crest than nuchal crest. Nuchal crest comprises low, overlapping, cresentic scales. Male does not have spine-like scales on crest. Dorsum greenish but varies, and may also be reddish-brown when stressed. Several radiating dark lines around eye orbit. Tail has eight dark bands; ventral area creamish in colour. **DISTRIBUTION** Found in southern Thailand and Peninsular Malaysia. **HABITS AND HABITAT** Arboreal and diurnal. Probably similar in habits to other *Gonocephalus* species. Occurs mainly in lowland forests, where strongly associated with larger trees.

◼ AGAMIDS ◼

Bell's Forest Dragon ◼ *Gonocephalus bellii* SVL 150mm

DESCRIPTION Typical species of *Gonocephalus* genus, with robust body and angle-headed head. Colour pattern varies between male and female but also with age. Male greyish-brown, female reddish-brown. Laterally covered in usually dark grey or black, net-like pattern. Orange tint to lateral sides seen in adult males in Perak (Malaysia). Gular pouch of male characteristic of species, with pale green colour at base and indigo-blue tip. Female gular has pink flecks. Tail has grey-brown and yellowish banding. Adult males more green in states of Kedah, Penang and Perak (Malaysia), compared to lizards from Terengganu, Selangor and Pahang. Hatchlings brown with reticulated pattern on body, black limbs and no colouration on dewlap. **DISTRIBUTION** Occurs thoughout Peninsular Malaysia, Singapore and southern Thailand, although records for Singapore have been questioned and it has been suggested that it has been brought into the country. **HABITS AND HABITAT** Diurnal and arboreal. Can be very cryptic in behaviour and difficult to see by day. Often found near or on large trees. At night, may be seen sleeping on branches. Inhabits lowland forests, where found to 100–1,100m asl.

AGAMIDS

Borneo Forest Dragon
Gonocephalus bornensis SVL 136mm

DESCRIPTION Very robust lizard with visible nuchal and dorsal crest – these are particularly developed in male. In female, nuchal crest higher than dorsal crest. Dorsum bright green with five dark bands. Lateral side has light oval spots. Tail light in colour with 12 dark bands. **DISTRIBUTION** Endemic to Borneo (Brunei Darussalam, Indonesia [Kalimantan], Malaysia [Sabah, Sarawak]). **HABITS AND HABITAT** Typical *Gonocephalus*: diurnal and arboreal. Often seen on tree trunks and saplings. Lives in tree trunks and also found on lianas near streams. Occurs in primary and secondary forests to 1,100m asl.

■ AGAMIDS ■

Chameleon Forest Dragon
■ *Gonocephalus chamaeleontinus* SVL 170mm

DESCRIPTION Robust body with dorsal crest comprising triangle-like scales. Head short and triangular in both dorsal and lateral profile. Curved eyebrows distinctive. Polymorphic species with both a green and a tan phase. Green phase: base colour of body, limbs and head green with darker radiating lines from eyes. Dark eye-patch sometimes seen. Reticulated pattern formed on body and head, with large yellow spots laterally. Tail black and white banded. Tan phase: tan base colour of body, head and limbs, with bolder reticulated pattern on lateral sides. Dark lines radiating from eye, which has a brown or yellow iris. Dewlap colour tan with thin darker stripes turning black distally. Tail black and white banded. **DISTRIBUTION** Found on west coast of Peninsular Malaysia (Pulau Tioman, Pahang State), and in Indonesia (Sumatra, Mentawai and Natuna archipelagos, and Java). **HABITS AND HABITAT** Diurnal and arboreal. Can be locally abundant. Clearly associates with trees. Often seen on sides of tree trunks; at night sleeps on branches. Inhabits lowland forests to about 500m asl.

◾ AGAMIDS ◾

Peter's Forest Dragon ◾ *Gonocephalus doriae* SVL 163mm

DESCRIPTION Brightly coloured, robust lizard with dorsal crest as high as nuchal crest. Both crests consist of low, overlapping scales. Male does not have spine-like scales on crests. Dorsum green with grey pattern or orange areas. Dorsum can change to reddish-brown with dark areas. Iris pink; yellow, orange or grey gular pouch with six distinctive bluish stripes. Flanks have eight transverse bars. Female and juveniles green with dark blotches. Ventral area creamish in colour. Limbs and tail pale green or grey with orange bands. **DISTRIBUTION** Endemic to Borneo, where found on both the Malaysian (Sabah and Sarawak) and Indonesian (Kalimantan) sides. Possibly also in Brunei Darussalam. **HABITS AND HABITAT** Diurnal and arboreal. Similarly to other *Gonocephalus* species, juveniles are most easily seen sleeping on leaves near streams or rivers. Particularly associated with low tree trunks and shrubs. Rather rare, and found in lowland rainforests, but also disturbed areas. Feeds primarily on arthropods.

▪ AGAMIDS ▪

Great Anglehead Lizard ▪ *Gonocephalus grandis* SVL 160mm

DESCRIPTION Large agama with large and triangular, but relatively long head. Colouration varies considerably between males and females (shown) and between adults and juveniles. Adults have green base colour with brown marbled markings; blue flanks with yellow spots; spots can extend to front legs. Legs green and black banded, with occasional yellow spots. Tail green black/brown striped from base. Adult male has large dorsal crest that is even larger on neck. Throat in both sexes fairly light in colour and may have some blue markings. Little geographic variation, although some females may have a pinkish hue across the body. Juveniles usually brown with darker brown markings and lighter throat. **DISTRIBUTION** Found in Indonesia (Kalimantan, Sumatra, Mentawai, Nias, Nako, Pinang and Tioman Islands), Laos, Malaysia, Singapore, Thailand and Vietnam. (Records from Laos and Vietnam need confirmation.) **HABITS AND HABITAT** Diurnal and arboreal. Mostly seen on trees or vegetation along streams. Sleeps on twigs and branches near streams by night. Juveniles usually seen closer to streams than adults. Will run or jump into the water when trying to escape from threats, and is a good swimmer. Can stay underwater for almost 15 minutes. Oviparous, laying several clutches of up to six eggs a year. Mainly found in peat swamp forests, thick lowland forests and hill dipterocarp forests below 1,400m asl; also does well in secondary forests.

Blue-eyed Anglehead Lizard ■ *Gonocephalus liogaster* SVL 140mm

DESCRIPTION Robust agama characterized by continuous nuchal and dorsal crests comprising lanceloate scales. In female nuchal crest is long, and dorsal crest low and ridge-like. Male brown or green, with reticulated darker pattern on flanks. Female has yellow cross-bars. Eye bright blue in male with reddish or orange skin surrounding orbit. Eye brown in female. **DISTRIBUTION** Found in Brunei Darussalam, Peninsular Malaysia, Borneo and Indonesia (Sumatra). Recently discovered in southern Thailand. **HABITS AND HABITAT** Diurnal and arboreal. Oviparous, with 1–4 eggs per clutch. Primarily found in lowland forests or peat swamp forests, where associated with tree trunks in vicinity of streams and small rivers.

◾ AGAMIDS ◾

Philippine Forest Dragon ◾ *Gonocephalus sophiae* TL 300mm

DESCRIPTION Typical *Gonocephalus* species. Nuchal crest well developed and much higher than dorsal crest. Dorsum brownish with green markings. Tail has black bands. Legs brown with green bands. Can have white spots on laterals; spots may also be green. **DISTRIBUTION** Endemic to the Philippines (Negros, Mindanao and Panay). Exact distribution unclear as it is often mistaken for **Boulenger's Forest Dragon** *G. interruptus* and the **Mindoro Forest Dragon** *G. semperi*. **HABITS AND HABITAT** Diurnal and arboreal. Found in primary and secondary lowland forests, often on tree trunks 1–2m above the ground and close to water. Quite common at 600m asl, but also found to 1,200m asl.

◾ Agamids ◾

Bornean Nose-horned Lizard ◾ *Harpesaurus borneensis* SVL 60mm

DESCRIPTION Lizards of this genus are characterized by a horn-like rostral appendage in either just the males, or in both sexes, of which the rostral appendage is more prominent on males. Body olive-green (male) or brown (female), with dark spots arranged diagonally. Low crest runs from neck to entire length of body. Long tail curls upwards into loose spiral when agitated, waggling around in a backwards and forwards locomotion similar to chameleons, the lizard making short leaps between twigs. The olive-green colour turns dull brown when stressed. **DISTRIBUTION** Found in Indonesia (Kalimantan), and Malaysia (Sarawak). **HABITS AND HABITAT** Lowland forest obligate; diurnal. Has been found at bases of limestone outcrops and near standing water or rocky streams in secondary forests. Sleeps in head-up position on tips of vines about 2–3.5m above the ground. In contrast to most agamids, which are generally oviparous, females of this species give birth to two live young per clutch (viviparity). Newly born lizard 26mm SVL; 70mm TL. Literature suggests that this species has switched to viviparity to adapt to local unpredictable floods.

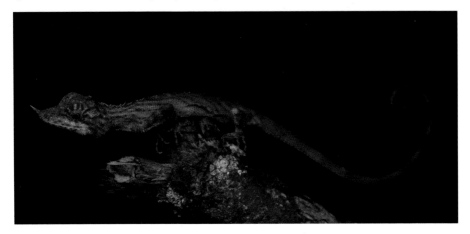

◾ AGAMIDS ◾

Amboina Sailfin Lizard ◾ *Hydrosaurus amboinensis* SVL 350mm

DESCRIPTION Very large lizard recognizable by large sailfin on tail; sailfin can reach to 12cm high in male, but slightly lower in female. Male also has distinctive nuchal and dorsal crest. Male black with orange-like/black spotting on lateral sides. Light blue iris. Sometimes difficult to distinguish from **Weber's Sailfin Lizard** *H. weberi*. Female more greyish than male, and juveniles green/black patterned. **DISTRIBUTION** Found in Indonesia (Ambon Island, eastern Maluku and New Guinea). Taxonomy of *Hydrosaurus* species in Indonesia still unresolved. **HABITS AND HABITAT** Sailfin lizards are diurnal and usually found in coastal forests adjacent to beaches, close to rivers, estuarine areas, deltas and river mouths in mangroves. They are linked to these riparian habitats, and can be found in high densities in suitable habitat.

▪ Agamids ▪

Philippine Sailfin Lizard ▪ *Hydrosaurus pustulatus* SVL 255mm

DESCRIPTION Large, brown to green sailfin lizard. High sail of up to 8cm on anterior part of tail, which is more prominent on male than female and used as a propeller when swimming. Crest runs along neck and dorsal area. Digits long and relatively large to aid movement on water. **DISTRIBUTION** Endemic to the Philippines. **HABITS AND HABITAT** Primarily diurnal and herbivorous. Strong swimmer that can run on the water's surface on its hindlegs for short distances when threatened. Oviparous, laying 2–8 eggs per clutch. Lives in various lowland habitats such as mangroves, coastal forests, banks of forest streams and rivers.

■ Agamids ■

Kinabalu Crested Dragon ■ *Hypsicalotes kinabaluensis* SVL 145mm

DESCRIPTION Very robust-looking lizard with well-developed nuchal crest and smaller dorsal crest. Characterized by subtympanic shield-like scale below tympanum. Male's gular poach brownish, and characterized by row of large, spiny scales that increase in size from jaw towards chest. Base colour green, with brown and black spots forming slight banding. **DISTRIBUTION** Only found in northern Borneo (Malaysia), on Gunung Kinabalu, Sabah. **HABITS AND HABITAT** Diurnal and arboreal. Inhabits submontane and montane forests. Occurs at 900–1,600m asl, on tree trunks and tree ferns.

Agamids

Beauty Butterfly Lizard
■ *Leiolepis belliana* SVL 170mm

DESCRIPTION Genus *Leiolepis* includes bisexual and unisexual (all-female) species. Head short and body robust and depressed. Male has ability to expand body to display colourful patterns on flanks for communication. Overall body grey to olive-brown. Dorsum has pale yellow, round or oval spots bordered with black, arranged in longitudinal rows. Spots may be connected to form three longitudinal stripes. Sides of body bright reddish-orange with 7–9 black vertical bars in male; 7–13 mid-body scale rows. Hindlegs have white oval and round spots bordered with black; 13–20 femoral pores on each side. Tail thick anteriorly and tapering posteriorly. Dorsal tail has light yellow, oval spots. Two subspecies currently recognized: *L. b. belliana* and *L. b. ocellata*. **DISTRIBUTION** Occurs in Cambodia, Indonesia, Malaysia, Myanmar, Thailand and Vietnam. **HABITS AND HABITAT** Diurnal and terrestrial. Lays 3–8 eggs per clutch. Inhabits open, flat areas in lowland rainforests.

■ Agamids ■

Spotted Butterfly Lizard
■ *Leiolepis guttata* SVL 184mm

DESCRIPTION Robust, slightly depressed-looking lizard with relatively small head compared to body. Dorsum pale grey or olive with pink spots and three dorsolateral stripes. Flanks blackish with seven vertical white bars. Forehead pale. Nape and dorsal surfaces of limbs and flanks of tail red or bright orange in male. Ventral area blue. **DISTRIBUTION** Endemic to southern Vietnam (Thua Thien Hue, Da Nang, Binh Dinh, Khanh Hoa, Ninh Thuan and Binh Thuan Provinces). **HABITS AND HABITAT** Diurnal and terrestrial. Lives in excavated burrows in loose, sandy soil. Feeds on flowers (crocus flowers) and arthropods. Reproductive ecology little known.

Reeves' Butterfly Lizard ■ *Leiolepis reevesii* SVL 170mm

DESCRIPTION Overall greyish-brown. Dorsum has bright, round orange-yellow spots with dark edges arranged close together, forming reticulation. Sides of body reddish-orange with black vertical bars in male. Sides of head have orange-and-black pattern. On each side, 13–16 femoral pores. Dorsal tail and hindlegs have small light spots. **DISTRIBUTION** Occurs in Cambodia, Thailand and Vietnam. **HABITS AND HABITAT** Diurnal and terrestrial. Feeds on vegetation and insects. Inhabits dry, flat and open areas.

■ AGAMIDS ■

Red-banded Butterfly Lizard ■ *Leiolepis rubritaeniata* SVL 170mm

DESCRIPTION Overall greyish-brown with highly reduced dorsal body pattern. Light yellow postocular stripe. Lateral body black with orange bars anteriorly and plain reddish-orange posteriorly in males. DISTRIBUTION Occurs in Cambodia, Laos, Thailand and Vietnam. HABITS AND HABITAT Diurnal and terrestrial. Builds burrow on the ground and often seen thermoregulating near burrow entrance. When disturbed, quickly retreats to burrow and covers entrance with loose soil. Inhabits flat, open areas in lowland rainforests and disturbed habitats.

Robinson's Anglehead Lizard
■ *Malayodracon robinsonii* SVL 152mm

DESCRIPTION Slender-looking lizard in which nuchal and dorsal crests are not separated, while slowly decreasing in height caudally. Large gular pouch pink in male and yellow in female. Overall dorsum green with dark cross-bars and yellow spots that can form bars. Labials white and separated by dark bars on sutures. Black postocular stripe visible towards tympanum. Tail banded. DISTRIBUTION Found in Peninsular Malaysia and extreme south of Thailand. HABITS AND HABITAT Diurnal and arboreal. Inhabits lowland rainforests. Ecology little known.

▪ AGAMIDS ▪

Phu Wua Lizard
▪ *Mantheyus phuwuanensis* SVL 90mm

DESCRIPTION Slender and slightly depressed-looking lizard with elongated snout. Male's gular pouch orange or yellow, and surrounded by two 'U'-shaped folds. Dorsum dark brown or olive-green with green spots. Flanks brown with black speckles and pale orange stripes. Ventral area blue-violet and yellow in male; unpatterned yellowish in female. **DISTRIBUTION** Found in west-central Laos (Bolikhamxay Province) and northeastern Thailand (Nong Khai Province). **HABITS AND HABITAT** Diurnal and nocturnal. Strongly associated with rocky habitats, where it can wedge itself in narrow rock crevices. Oviparous, laying up to four eggs per clutch, which are stuck in rock crevices. Inhabits rocky areas in semi-evergreen lowland forests, primarily at 200–300m asl.

Black-lipped Shrub Lizard ▪ *Pelturagonia nigrilabis* SVL 58mm

DESCRIPTION Short, robust-looking lizard that lacks spine above eye. Nuchal crest comprises 6–12 scales. Gular scales keeled. Dorsum brown, green or olive, without transverse blue banding in male. Females brown to olive. Ventral area creamish with dark brown patterns. Inner lining of mouth blue or blackish. **DISTRIBUTION** Occurs in Borneo (Malaysia and Indonesia), and on Pulau Sirhassen in Natuna archipelago (Indonesia). **HABITS AND HABITAT** Diurnal and arboreal. When threatened, reveals blue or black inner lining of mouth. Strongly associated with shrubs and lower parts of tree trunks. Found in dipterocarp forests.

◾ Agamids ◾

Chinese Water Dragon ◾ *Physignathus cocincinus* SVL 250mm

DESCRIPTION Base colour dark to light green with diagonal stripes of green or turquoise on body. Tail banded from middle to end with light green and black. Head green and triangular, with white or purple lower jaw and throat. Male has large head and develops large crest on head, neck, dorsum and tail. Female generally smaller than male. DISTRIBUTION Found in east and south-east Thailand, Cambodia, Vietnam, Laos and Myanmar. HABITS AND HABITAT Diurnal. Often seen basking in trees or other plants. Good swimmer, and can remain submerged when threatened. Oviparous, laying 5–15 eggs per clutch. Native to lowland and highland forests, and occurs most commonly on banks of freshwater streams and lakes.

Yellow-throated False Garden Lizard
◾ *Pseudocalotes flavigula* SVL 72mm

DESCRIPTION Relatively robust body with olive-green dorsum, changeable to brownish colour. Small nuchal crest of six spiny scales. Dorsal crest less developed. Saddle-like brown patches. Gular pouch bright yellow in male; mustard-yellowish in female. Forehead greenish with dark speckles. Throat and lips white. Tail has dark bands. DISTRIBUTION Found in Peninsular Malaysia (Pahang State). Originally described from Cameron Highlands (Pahang). HABITS AND HABITAT Diurnal and arboreal. Often found on or near tall trees. Inhabits montane forests at about 1,520–2,012m asl.

◾ AGAMIDS ◾

Flower's Forest Agamid ◾ *Pseudocalotes floweri* SVL 98mm

DESCRIPTION Slender, slightly compressed-looking agamid with long head. Nuchal crest has eight spines. Dorsum brown with enlarged blue dorsal scales. Spines light brown. Eyes have radiating darker brown lines extending to supraorbital ridge. Limbs have irregular dark brown bands. Gular has dark brown spot. Tail has up to 10 dark brown spots. **DISTRIBUTION** Found in south-east Thailand (Khao Sebab, Chantaburi Province), Cambodia (Cardamom Mountains). **HABITS AND HABITAT** Diurnal and arboreal. Found in submontane or secondary forests at 1,200–2,134m asl.

Burmese Mountain Agamid ◾ *Pseudocalotes kakhienensis* SVL 125mm

DESCRIPTION Robust-looking agamid with large, broad head. Gular pouch absent. Nuchal crest comprises 7–9 spines. Irregular small, keeled scales with groups of larger keeled scales on dorsal area. Dorsum pale greenish or olive, with light and dark brown markings. Ventral area green, speckled with dark brown streaks. Postocular stripe to tympanum. Supralabials and infralabials have dark bars on sutures. **DISTRIBUTION** Found in Myanmar (east of Ayeyarwady river), northern Thailand and southern China (Yunnan Province). **HABITS AND HABITAT** Diurnal and arboreal. Oviparous, with clutches of up to five eggs. Occurs in submontane forests at 1,200–1,400m asl.

◾ Agamids ◾

Khao Nan Long-headed Lizard
◾ *Pseudocalotes khaonanensis* SVL 104.5mm

DESCRIPTION Largest of the *Pseudocalotes*; robust looking. Gulars distinct in male, and nuchal crest comprises nine spines connected to indistinct dorsal crest. Dorsum rust-brown with irregularly arranged darker or lighter scales and three darker bands. Dark rust-brown lines radiating from eye. Gular pouch red with cream to white scales. Labials white with black banding on sutures. Dark postocular stripe towards tympanum. Tail dark banded. **DISTRIBUTION** Endemic to Thailand, where known from Khao Nan National Park. **HABITS AND HABITAT** Arboreal and diurnal. Found on tree growth below 4m tall. Inhabits montane scrub cloud forests above 1,100m asl.

Bukit Larut False Garden Lizard
◾ *Pseudocalotes larutensis* SVL 77mm

DESCRIPTION Relatively slender lizard with small nuchal crest of five spiny scales. Pale yellowish base colour with greenish cast. Dorsum has three broad dark bands. Gular pouch has yellow spot, bordered by two dark spots. Tail banded with reddish-brown colour. **DISTRIBUTION** Endemic to Peninsular Malaysia (Bukit Larut, Perak State). **HABITS AND HABITAT** Diurnal and arboreal. Associated with trees and often seen at 7–15m above the ground. Mostly found in submontane forests at about 1,000m asl.

◾ AGAMIDS ◾

Small-scaled Montane Forest Lizard
◾ *Pseudocalotes microlepis* SVL 85mm

DESCRIPTION Distinctive long, projecting snout and moderately compressed body. Head and body greyish-brown to brown dorsally and lighter ventrally with dark spots. Nuchal crest a series of compressed erect scales. Dorsal crest very low and pointing backwards. Scales on upper body heavily keeled; 71–74 scales around mid-body; 7–8 upper and 7–9 lower labials; 18–19 lamellae beneath fourth finger; 21–25 lamellae beneath fourth toe. Tail long, about two times longer than SVL. **DISTRIBUTION** Occurs in Laos, Myanmar, Thailand and Vietnam. **HABITS AND HABITAT** Diurnal and arboreal. Found on vegetation or rocks at about 1m above the ground. Inhabits montane forests above 1,100m asl.

■ AGAMIDS ■

Green Fan-throated Lizard ■ *Ptyctolaemus gularis* SVL 80mm

DESCRIPTION Slender, slightly compressed-looking lizard. Dorsal scales keeled and dorsal crest absent. Three longitudinal folds on each side of throat meet on back. Gular pouch present; femoral pores absent. Dorsum olive-brown with fold on back deep blue in colour. Five broad bands visible on body; sometimes green dorsolateral band on front of body. Flanks have network of darker brown and rounded flecks of green. **DISTRIBUTION** Found in Myanmar (Kachin State, Sagaing Division and Chin State). Also India, Bangladesh and Tibet. **HABITS AND HABITAT** Diurnal and arboreal. Found in urban habitats and submontane forests at 182–1,220m asl.

Glass Lizards

Anguidae (Glass Lizards)

The family of glass lizards, which resemble snakes, in fact comprises limbless lizards. Anguidae are heavy-bodied species that have no overlapping scales underlaid by osteoderms. They are generally diurnal in nature, terrestrial or semi-fossorial, and can often be found in open areas and forests. Even though they look like snakes, their tails are capable of autotomy.

Burmese Glass Lizard ■ *Dopasia gracilis* SVL 208mm

DESCRIPTION Large, robust glass lizard. Dorsum can be reddish-brown or yellowish, with several rows of blue spots edged with dark scales. Iris yellowish. Easily mistaken for **Buettikofer's Glass Lizard** *D. buettikoferi*, but range does not overlap. **DISTRIBUTION** Found in Myanmar, northern and northeastern Thailand, Laos and Vietnam. Also India, Bangladesh and southern China. **HABITS AND HABITAT** Diurnal and terrestrial. Plays dead when feeling threatened. Inhabits submontane and montane forests.

■ BLIND LIZARDS ■

DIBAMIDAE (BLIND LIZARDS)
These snake-like lizards have small. close-fitting, glossy scales and completely lack forelimbs. They do, however, have vestigial hindlegs that have been reduced to scaly flaps. Blind lizards can autotomize their tails when threatened. Secretive and fossorial in nature, they are rarely found on the surface but more often under logs.

Boo Liat's Worm Lizard ■ *Dibamus booliati* SVL 102.7mm
DESCRIPTION Legless lizard with brownish-red base colour. Similarly to the **White-tailed Worm Lizard** *D. bourreti*, every scale seems to be dark edged. Snout-tip pale, and creamish collar at nape. Female can lack hindlimb flaps characteristic of *Dibamus* males.
DISTRIBUTION Found in Peninsular Malaysia (Batu Gua Madu, Kelantan State).
HABITS AND HABITAT Fossorial. Burrows in limestone rubble and often found under debris. When threatened, will pretend to be dead. Found in secondary forests in limestone areas to 121m asl.

■ Blind Lizards ■

Flower's Blind Lizard ■ *Dibamus floweri* SVL 112mm

DESCRIPTION Worm-like and almost cylindrical legless lizard. Dorsum, flanks and tail grey-brownish, with silver band midway down body of about 10–12 scales in length. Labial and nasal scales beige to opaque. Ventral area lighter in colour. Female lacks hindlimbs that are characteristic of *Dibamus* males. Body of holotype is merely 3.55mm wide, with tail only slightly narrower at 3.11mm wide. Can be distinguished from Boo Liat's Worm Lizard (p. 51) and **Tioman Worm lizard** *D. tiomanensis* due to absence of labial and nasal sutures in Flower's Blind Lizard. In addition to differences in scale counts, there is no incomplete rostral suture in Boo Liat's and Tioman Worm Lizards, present in Flower's Blind Lizard. **DISTRIBUTION** Only known from Fraser's Hill, Pahang, Peninsular Malaysia. **HABITS AND HABITAT** Can be seen digging through leaf litter along banks of roads and has been found hiding beneath soil by day. Flares up body scales as defensive mechanism when disturbed, typical of *Dibamus* species, and most likely mimics non-palatable species of worm. Montane species, currently only known to occur at 1,207–1,500m asl.

▪ Eyelid Geckos ▪

Eublepharidae (Eyelid Geckos)
This family of eyelid geckos is found in Asia, North America and Africa. Species in the family have fleshy eyelids and very soft skin. In contrast to most other geckos, they can blink, and they have well-developed limbs with narrow digits. They encompass a wide range of species, including those that are arboreal, terrestrial and saxicolous.

Cat Gecko ▪ *Aeluroscalabotes felinus* SVL 122mm

DESCRIPTION Tan-brown lizard with dark or pale vertebral stripe on body and tail. Tail-tip can be white. Labials pale/white, showing strong contrast to rest of body. Juveniles more brightly coloured than adults. Feet have lamellae, but only on bases of digits. Tail often curled up. **DISTRIBUTION** Found in Thailand, Peninsular Malaysia (and Sarawak), Singapore and Indonesia (Sumatra and Kalimantan). **HABITS AND HABITAT** Nocturnal and arboreal. Often found in low vegetation or in or around dead logs. Inhabits lowland rainforests and peat swamp forests to about 1,000m asl.

▪ Eyelid Geckos ▪

Vietnamese Leopard Gecko ▪ *Goniurosaurus araneus* SVL 124mm

DESCRIPTION Gracile gecko characterized by nuchal loop and four thick, full-body bands between forelimbs and caudal constriction, which are pale to bright orange. Five white caudal bands. Dorsum lacks mottling and is dull yellow or greyish. Head can contain dark markings or spots. **DISTRIBUTION** Found in northern Vietnam (Cao Bang Province) and China (Guangxi). **HABITS AND HABITAT** Nocturnal. Most active on rocky surfaces, but can occasionally be found on forest floors. Often located on north-facing slopes. Occurs in relatively dry rocky or karst areas covered with thick tropical rainforests. Seen at 180–260m asl.

Cat Ba Leopard Gecko ▪ *Goniurosaurus catbaensis* SVL 111.5mm

DESCRIPTION Gracile gecko characterized by 3–4 thin dorsal body-bands, surrounded by dark narrow band between limbs. Ground colour grey-brown to pale brown and mottled. Head pattern has dark marbling. Iris orange-brown. Tail-base characterized by another dorsal body-band; tail-base grey-brown to dark brown, with five white caudal bands. **DISTRIBUTION** Endemic to Cat Ba Island, Vietnam. **HABITS AND HABITAT** Nocturnal. Oviparous, laying two eggs per clutch, to six times a year. Mainly found on the forest floor at night, but also to 2–3m on limestone cliffs. Mostly occurs near large limestone caves with forest vegetation, or near shrub vegetation on limestone.

▪ Eyelid Geckos ▪

Lichtenfelder's Gecko ▪ *Goniurosaurus lichtenfelderi* SVL 83mm

DESCRIPTION Relatively robust-looking gecko with dark or light brown dorsum. Two dark-edged dorsal bands between front legs and hindlimbs. Ventral area pale grey. **DISTRIBUTION** Found in northeastern Vietnam (Lang Son, Quang Ninh, Bac Giang, Ha Bac and Hai Hung Provinces). Also islands in Gulf of Tonkin. **HABITS AND HABITAT** Nocturnal and saxicolous. Strong association with granite-dominated rocky streams. Inhabits lowland forests at 100–600m asl.

Chinese Leopard Gecko ▪ *Goniurosaurus luii* SVL 119mm

DESCRIPTION Slender and gracile gecko with grey-brown to white base colour, mottled with small dark brown blotches. Iris orange. Three thin yellowish bands edged by dark bands make up dorsal. Tail black with five thin white caudal bands. **DISTRIBUTION** Found in Cao Bang Province, Vietnam, and in Guangxi and probably Hainan Provinces, southeastern China. **HABITS AND HABITAT** Nocturnal. Mainly active during the wet months (June–August), and found up to 5m into limestone caves. Oviparous, laying two eggs per clutch, to six times a year. Occurs in tropical secondary forests in rocky areas near limestone caves to 770m asl.

Typical Geckos

Gekkonidae (Typical Geckos)

This is one of the largest lizard families, with 473 recognized species in Southeast Asia at the time of writing, and new species being described regularly. These geckos inhabit a wide range of habitats. All species are oviparous, laying just one or two hard-shelled eggs, often with multiple clutches a year.

Bauer's Rock Gecko ■ Cnemaspis baueri SVL 64.9mm

DESCRIPTION Robust, elongated-looking gecko with large, depressed head. Dorsum surface has enlarged tubercles that lack white tips. Precloacal and femoral pores in male. Dorsum dark brown to olive, with dark grey spots on vertebrals. Forehead has dark spots and two broken lines, radiating from orbit to back of head. Limbs and tail not dark banded. Iris orange. Ventral area brownish. **DISTRIBUTION** Found in Pulau Tulai, Johor, Peninsular Malaysia. **HABITS AND HABITAT** Diurnal and saxicolous. Strongly associated with rocks and boulders. Found in lowland forests.

■ Typical Geckos ■

Twin-spotted Rock Gecko ■ *Cnemaspis biocellata* SVL 40mm

DESCRIPTION Characterized by slender, elongated body, with flattened head distinct from neck, large, forwards- and upwards-looking eyes, long, widely splayed limbs, and long digits. Overall colour yellowish-brown (male) to greyish-brown (female) with light vertebral blotches. Two distinct white occipital ocelli. Small black shoulder-blotch encircling white or yellow round spot on each side. Labials: 6–10 upper and 5–9 lower. Lamellae: 29–37 beneath fourth toe. Tail same colour as body, with light blotches. **DISTRIBUTION** Found in Malaysia and Thailand. **HABITS AND HABITAT** Diurnal and arboreal. Very agile and wary; retreats to crevices when threatened. Occurs on karst formations and rocks in lowland forests.

Chan-ard's Rock Gecko ■ *Cnemaspis chanardi* SVL 40mm

DESCRIPTION Dorsum overall brown with dark markings on neck and back. Dorsal body has light vertebral spots. Large tubercles on body and tail. Sides of body have thin white or yellow vertical bars. Labials: 7–10 upper and 6–8 lower. Lamellae: 25–30 beneath fourth toe. Precloacal pores: 6–8 separated in middle by non-pore-bearing scales in male. Tail has series of irregular-shaped, thick dark and broader light spots. **DISTRIBUTION** Found in Thailand. **HABITS AND HABITAT** Diurnal and arboreal. Found on tree buttresses or near tree holes 1.5–2.0m above the ground and near rock outcrops. Occurs from near sea level to 600m asl in rainforests.

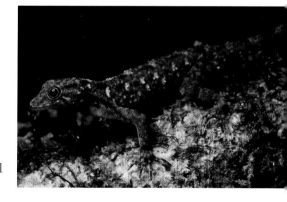

▪ Typical Geckos ▪

Chanthaburi Rock Gecko ▪ Cnemaspis chanthaburiensis SVL 42mm

DESCRIPTION Overall brown to greyish-brown. Head and tail may be yellowish-brown. Labials: 8–10 upper and 7–10 lower. Lamellae: 22–29 beneath fourth toe. Dorsal body has dark and light oval markings. Sides of body yellow or orange anteriorly. Throat, abdomen and subcaudal orange. Tail has alternating light brown and yellowish or whitish transverse bands. Precloacal pores: 6–9. Postcloacal tubercles: 1–3 on each side. DISTRIBUTION Found in Cambodia and Thailand. HABITS AND HABITAT Nocturnal and terrestrial. Commonly found hiding by day in loose bark and logs on forest floors. Occurs from near sea level to 970m asl.

Kendall's Rock Gecko ▪ Cnemaspis kendallii SVL 58mm

DESCRIPTION Overall grey to dark brown. Dorsal head has dark and light greyish-brown markings. Two postocular stripes; 10–11 upper and 8–9 lower labials. Dorsal body has rows of moderate dark and light brown, round and oval spots, and small off-white or yellow spots. Limbs and digits banded; 25–33 lamellae beneath fourth toe. No precloacal pores. Two postcloacal tubercles on each side. Tail has alternating dark and light bands. Subcaudal white. DISTRIBUTION Found in Indonesia (Kalimantan, Borneo), Malaysia (Sarawak, Borneo) and Singapore. HABITS AND HABITAT Diurnal and arboreal. Commonly found on shaded areas of large granite boulders, tree trunks and roots, and limestone formations. Inhabits primary and secondary rainforests.

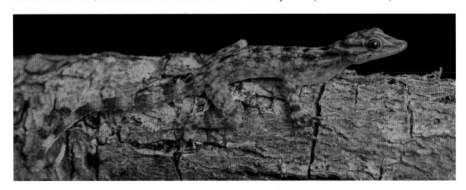

▪ Typical Geckos ▪

Kumpol's Rock Gecko ▪ Cnemaspis kumpoli SVL 60mm

DESCRIPTION Sexually dimorphic, with male more colourful than female. Male: overall colour yellowish-green to yellow with red blotches on head, body, limbs and tail. Pair of black shoulder-patches enclosing longitudinal white bar on each. Female and juveniles: greyish-brown with paired dark marking on dorsal body followed by light blotches. Labials: 7–9 upper and 6–8 lower. Lamellae: 34–41 beneath fourth toe. Precloacal pores: 1–8 discontiguous. Precloacal tubercles: 2–3 on each side. **DISTRIBUTION** Occurs in Malaysia and Thailand. **HABITS AND HABITAT** Nocturnal and arboreal. Female lays two eggs per clutch. Inhabits primary lowland forests and commonly seen on granite rocks.

Tioman Island Rock Gecko
▪ Cnemaspis limi SVL 88mm

DESCRIPTION Overall dark brown with thin yellow and dark markings on head and body. Two postocular stripes; 8–12 upper and 7–10 lower labials. Dorsal body has 5–7 paired dark oval spots. Lamellae: 29–36 beneath fourth toe. Dorsal tail has white tubercles arranged transversely. **DISTRIBUTION** Occurs in Malaysia. **HABITS AND HABITAT** Diurnal and arboreal. Mainly found on large granite boulders in head-down position. Inhabits primary and secondary rainforests from sea level to 1,026m asl. Sympatric with the **Peninsular Rock Gecko** C. peninsulari.

▪ Typical Geckos ▪

McGuire's Rock Gecko ▪ *Cnemaspis mcguirei* SVL 65mm

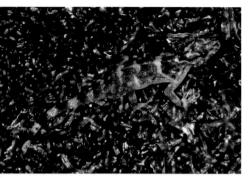

DESCRIPTION Overall grey to brown with irregularly shaped light and dark markings. Labials: 7–10 upper and 7–9 lower. Dark patch on shoulder enclosing two yellow ocelli. Pair of bands behind front limbs. Light roundish markings from neck to tail-base. Bright yellow markings on sides of body. Lamellae: 27–35 beneath fourth toe. Precloacal pores: 5–10 that may be discontiguous or continuous; 2–3 postcloacal spurs. Tail has brown and dull white bands. **DISTRIBUTION** Occurs in Peninsular Malaysia and Thailand. **HABITS AND HABITAT** Diurnal and arboreal. Usually seen on granite rocks without moss cover. Inhabits lower montane forests at 800–1,350m asl.

Fairy Rock Gecko ▪ *Cnemaspis paripari* SVL 51mm

DESCRIPTION Overall brown (female) or yellowish-brown (male). Head bright yellow in adult male. Dark postocular stripe; 12–13 upper and 10–11 lower labials. Dorsal body has thin yellow transverse bands. Limbs same colour as body, usually without markings. Digits long and slender, lighter in colour; 26–31 lamellae beneath fourth toe. Tail grey anteriorly and white (original) or bright yellow (regenerated) posteriorly in adult male; brown anteriorly and grey posteriorly in female, without distinct bands. Subcaudal has median row of bead-like scales. **DISTRIBUTION** Occurs in Indonesia and Malaysia. **HABITS AND HABITAT** Diurnal and arboreal. Usually found on vertical surfaces of rocks and in shaded areas. Retreats to rock crevices when disturbed. Only known to occur in karst outcrops.

■ Typical Geckos ■

Peninsular Rock Gecko ■ *Cnemaspis peninsularis* SVL 60mm

DESCRIPTION Overall brown to yellowish-brown; cream with faint markings when at rest. Dorsal body has 2–3 longitudinal rows of oval spots. Dark spots may be followed by light oval spots. Labials: 10–11 upper and 7–10 lower. Lamellae: 27–33 beneath fourth toe. Banded digits. Adult male lacks precloacal and femoral pores; 1–2 precloacal tubercles on each side. Banded tail. Regenerated tail yellow in male. **DISTRIBUTION** Occurs in Malaysia and Singapore. **HABITS AND HABITAT** Diurnal and arboreal. Commonly seen on vegetation, logs, tree trunks and rocks. Male curls tail forwards and moves it from side to side when threatened, before retreating to crevice. Inhabits primary and secondary forests from sea level to 260m asl.

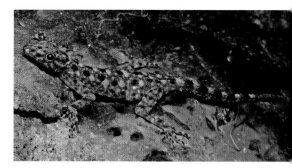

Phuket Rock Gecko ■ *Cnemaspis phuketensis* SVL 29mm

DESCRIPTION Small, overall brown to olive-brown rock gecko. Thin black stripe from snout through eye to postocular; 6–7 upper and 6–7 lower labials. Dorsal body has dark markings and light vertebral spots. Lamellae: 16–17 beneath fourth toe. Adult male lacks precloacal and femoral pores. Tail length 1.2 times SVL. **DISTRIBUTION** Occurs in Thailand. **HABITS AND HABITAT** Diurnal and arboreal. Seen on vegetation and tree trunks less than 1m above the ground and on banks of streams.

■ TYPICAL GECKOS ■

Psychedelic Rock Gecko ■ *Cnemaspis psychedelica* SVL 75mm

DESCRIPTION The most unusually coloured rock gecko, with bright orange legs, feet and tail. Head dirty yellow anteriorly and with yellow reticulated pattern over black blotches posteriorly; 7–10 upper and 5–8 lower labials. Dorsal body blue-grey or light purple. Sides of body bright orange with thin yellow vertical bars. Lamellae: 24–28 beneath fourth toe. **DISTRIBUTION** Occurs in Vietnam. **HABITS AND HABITAT** Arboreal and diurnal, but may also be active at night. Female lays two eggs per clutch and adheres them to flat rock surfaces. Type locality is an offshore island in south Vietnam. Preferred microhabitat is granite rocks and cliffs with canopy cover in lowland rainforests. **NOTE** Threatened by illegal wildlife trade due to its unique colouration. Listed in CITES Appendix I since 2017, prohibiting the international trade of wild-caught individuals for commercial purposes.

Van Deventer's Rock Gecko ■ *Cnemaspis vandeventeri* SVL 45mm

DESCRIPTION Head oval and narrow when viewed from above, distinct from neck. Dorsal head with small brown markings. Faint dark stripe behind eye; 8–9 upper and 7–9 lower labials. Dorsal body has small tubercles, absent on lower sides of body. Vertebral blotches from nape to hindlimbs. No dark markings or blotches on throat. Limbs mottled; digits with dark bands; 24–28 lamellae beneath fourth toe. Four precloacal pores separated in middle by non-pore-bearing scales in male; 1–3 postcloacal tubercles. **DISTRIBUTION** Occurs in Thailand. **HABITS AND HABITAT** Arboreal and nocturnal. Very limited information on natural history. Individuals were found on vines along a stream and beneath decaying wood.

▪ TYPICAL GECKOS ▪

Agusan Bent-toed Gecko ▪ *Cyrtodactylus agusanensis* SVL 100mm

DESCRIPTION Dorsal body light grey to brown with dark, irregularly shaped transverse bands and relatively large conical or pointed tubercles. Labials: 7–10 upper and 6–8 lower. Lamellae: 23–30 beneath fourth toe. Precloacal pores: 6–9 arranged in inverted 'V'. Femoral pore-bearing scales: 3–11, separated medially by non-pore-bearing scales. Postcloacal tubercles: 4–7 on each side. **DISTRIBUTION** Occurs in the Philippines (Mindanao). **HABITS AND HABITAT** Nocturnal and terrestrial. Commonly seen beneath rotten logs and rocks along river and stream banks. Found in riparian habitats near streams and rivers at 400–1,000m asl.

Annulated Bent-toed Gecko ▪ *Cyrtodactylus annulatus* SVL 80mm

DESCRIPTION Dorsum light greyish-brown overall, with 4–5 irregularly shaped, dark transverse markings and conical or pointed tubercles. Labials: 7–10 upper and 7–8 lower. Dark brown stripe from snout through eye to nape, and dark 'M'-marking on neck. Limbs and digits banded; 17–22 lamellae beneath fourth toe. Six precloacal pores in male and femoral pore-bearing scales lacking; 2–4 postcloacal tubercles on each side. Tail has light and dark transverse bands; dark bands on original tail broader than light bands posteriorly. **DISTRIBUTION** Found in the Philippines. **HABITS AND HABITAT** Nocturnal and arboreal. Commonly found on tree trunks and under rotting logs on forest floors. Female lays two eggs per clutch. Hatchling 25–28mm SVL. As currently understood, occurs in Visayas and Mindanao faunal regions at sea level to 1,200m asl in riparian habitats near streams.

■ TYPICAL GECKOS ■

Southern Titiwangsa Bent-toed Gecko
■ *Cyrtodactylus australotitiwangsaensis* SVL 120mm

DESCRIPTION Dorsum dark beige overall. Labials: 9–12 upper and 9–13 lower. Nuchal band extends from postocular; bordered with thin, cream-coloured margins.

Dorsal body has 3–4 dark brown transverse bands bordered by white tubercles, forming dotted lines; 22–30 longitudinal rows of dorsal tubercles; 39–45 contiguous femoral-precloacal pore-bearing scales in male. Original tail has alternating dark and immaculate white transverse bands. **DISTRIBUTION** Found in Malaysia. **HABITS AND HABITAT** Nocturnal and arboreal. Occurs on southern portion of Titiwangsa Mountains at 1,000–1,500m asl.

Balu Bent-toed Gecko ■ *Cyrtodactylus baluensis* SVL 95mm

DESCRIPTION Dorsum light brown, with or without dark blotches on head. Tubercles on neck, dorsum, limbs and tail prominent. Dorsum has irregularly shaped, brownish-black transverse bands.

Limbs and digits banded; 19–23 lamellae beneath fourth toe; 9–10 precloacal and 6–9 femoral pores in male. Original tail has dark and light transverse bands. **DISTRIBUTION** Occurs in Brunei Darussalam and Malaysia (Borneo). **HABITS AND HABITAT** Nocturnal and arboreal. Usually seen on tree trunks and bases of trees. Female lays two eggs per clutch. Hatchling 32mm SVL; 68mm TL. Inhabits montane forests at 900–2,200m asl.

Bintang Mountain Bent-toed Gecko
■ *Cyrtodactylus bintangtinggi* SVL 111mm

DESCRIPTION Member of the **Malayan Bent-toed Gecko** C. *pulchellus* complex. Dorsal head and body brown. Labials: 9–13 upper and 8–11 lower. Nuchal band extends from behind eye. Dorsum has 3–4 dark brown transverse bands bordered by white tubercles, and 22–25 longitudinal rows of dorsal tubercles; 20–24 lamellae beneath fourth toe; 39–41 contiguous femoral precloacal pore-bearing scales in male. Original tail has 8–10 dark brown bands. White caudal bands immaculate. Hatchling dorsum pale yellow with dark bands; tail has alternating black and immaculate white transverse bands. **DISTRIBUTION** Occurs in Malaysia and Thailand. **HABITS AND HABITAT** Nocturnal and arboreal. Inhabits upland forests to 1,500m asl. Usually found on and near vicinity of granite rocks.

Short-handed Bent-toed Gecko
■ *Cyrtodactylus brevipalmatus* SVL 73mm

DESCRIPTION Cryptic-coloured bent-toed gecko with silver-white iris and prehensile tail. Dorsum brown with dark brown marking on nape and body. Labials: 12–13 upper and 10–11 lower. Tubercles: 12–18 at mid-body. Webbing at bases of digits; 16–19 lamellae beneath fourth toe. Precloacal pores: 9–10 in male; 6–7 femoral pores. Tail square in cross-section, usually kept tightly coiled and close to body. **DISTRIBUTION** Occurs in Thailand. **HABITS AND HABITAT** Nocturnal and arboreal. Individuals have been found on trees and hiding beneath dead bark. Inhabits rainforests at about 750m asl.

▪ Typical Geckos ▪

Banded Bent-toed Gecko ▪ *Cyrtodactylus consobrinus* SVL 125mm

DESCRIPTION Large bent-toed gecko with brownish-black dorsum. Snout has yellowish narrow lines and dots; 10–16 upper and 9–13 lower labials. Top of head has narrow network of lines. Dorsum body and limbs have narrow yellowish or white transverse bands; 22–28 lamellae beneath fourth toe; 9–10 precloacal and 0–6 femoral pores in male. Tail moderate, tapering to a point and with narrow white bands. **DISTRIBUTION** Occurs in Brunei Darussalam, Indonesia, Malaysia, Singapore and Thailand. **HABITS AND HABITAT** Nocturnal and arboreal. Found on tree trunks and in crevices. Inhabits lowland rainforests.

Cryptic Bent-toed Gecko ▪ *Cyrtodactylus cryptus* SVL 91mm

DESCRIPTION Moderate-sized bent-toed gecko. Dorsum light brown to light greyish-lavender. Snout and dorsal head have dark oval blotches. Dark brown nuchal loop extends to postocular. Dorsal body has irregularly shaped transverse bands bordered by narrow yellowish margins; 19–20 longitudinal rows of dorsal tubercles. Limbs and digits banded; 20–23 lamellae beneath fourth toe; 9–11 precloacal pores in male and 16–27 enlarged precloacal scales in both sexes. No enlarged femoral scales. Tail has alternating dark brown and cream transverse bands. Median subcaudal scales not enlarged. **DISTRIBUTION** Found in Laos and Vietnam. **HABITS AND HABITAT** Nocturnal and arboreal. Female lays two hard-shelled, oval eggs per clutch, measuring 12 x 14mm. Occurs in karst forests and usually seen to 2m above the ground on karst cliffs, tree trunks and branches and other vegetation; occasionally on the ground.

▪ Typical Geckos ▪

Doi Suthep Bent-toed Gecko
▪ *Cyrtodactylus doisuthep* SVL 90mm

DESCRIPTION Dorsum blackish-brown overall, with yellowish reticulated pattern on top of head; 10–12 upper and 8–11 lower labials. Dorsal body has irregularly shaped, narrow beige transverse bands and 19–20 longitudinal rows of tubercles. Sides of body have prominent white tubercles. Ventral uniform beige. Both sexes have 34–35 enlarged femoral-precloacal scales. Six precloacal pores in male. Tail has yellowish bands posteriorly and whitish bands anteriorly. DISTRIBUTION Occurs in Thailand. HABITS AND HABITAT Nocturnal and terrestrial. Usually found on rocks near streams and exposed tree roots, and beneath logs. Inhabits forests at 350–1,660m asl.

White-eyed Bent-toed Gecko ▪ *Cyrtodactylus elok* SVL 68mm

DESCRIPTION Dorsum light reddish-brown with triangular-shaped dark brown marking on nape. Iris silver-white. Narrow whitish postocular stripe. Dorsal body has interrupted brown transverse bands; 6–10 tubercles at mid-body; no tubercles on sides of body. Digits have webbing at bases; 18–19 lamellae beneath fourth toe. Eight precloacal and no femoral pores in male; 8–10 enlarged precloacal scales in both sexes. Prehensile tail square in cross-section, with extensive caudal fringe, usually kept tightly coiled and close to body.
DISTRIBUTION Occurs in Malaysia and Thailand.
HABITS AND HABITAT Nocturnal and arboreal. Has been seen to move slowly in stems with the aid of its prehensile tail. Holotype was found in leaf litter on the forest floor at 215m asl.

▪ Typical Geckos ▪

Phetchabun Bent-toed Gecko
▪ *Cyrtodactylus interdigitalis* SVL 80mm

DESCRIPTION Dorsum light yellowish-brown to brown overall. Labials: 10–13 upper and 10–12 lower. Brown nuchal loop extends to posterior of eye. Dorsal body has irregularly shaped brown transverse bands and 16–22 longitudinal rows of tubercles. Edges of dark bands and light bands have black spots; 14–16 precloacal pores and 7–9 femoral pores on each thigh in male. Tail has dark transverse bands; prehensile. DISTRIBUTION Occurs in Laos and Thailand (Phetchabun and Loei Provinces). HABITS AND HABITAT Nocturnal and arboreal. Usually found in and around caves and tree trunks. Reported to occur in Laos, but identity needs to be re-evaluated.

Intermediate Bent-toed Gecko
▪ *Cyrtodactylus intermedius* SVL 79mm

DESCRIPTION Dorsum greyish-brown overall. Nuchal loop and four dorsal body-bands bordered by thin yellowish margins. Eight upper and 9–10 lower labials; 19–20 longitudinal rows of dorsal tubercles at mid-body; 19–21 lamellae beneath fourth toe; 8–10 precloacal pores in male; 23–24 enlarged femoral scales. Tail has alternating dark brown and dirty white transverse bands. DISTRIBUTION Found west of Cardamom Mountains in southeast Thailand. HABITS AND HABITAT Nocturnal and arboreal. Often seen on tree trunks and boulders near streams. Inhabits forests at 50–1,000m asl.

▪ Typical Geckos ▪

Jarujin's Bent-toed Gecko ▪ *Cyrtodactylus jarujini* SVL 90mm

DESCRIPTION Dorsum light brown with a few spots on top of head and irregular blotches on body. Nuchal loop interrupted medially and extends to postocular; 18–20 longitudinal rows of dorsal tubercles at mid-body. Limbs and digits have dark brown blotches. Ventral uniform whitish. Tail relatively narrow, tapering to a point, with alternating dark and light brown transverse bands. DISTRIBUTION Occurs in Laos and Thailand. HABITS AND HABITAT Nocturnal and arboreal. Often seen on steep sides of boulders. Hides behind vegetation around boulders when disturbed. Inhabits sandstone hills.

Lekagul's Bent-toed Gecko ▪ *Cyrtodactylus lekaguli* SVL 104mm

DESCRIPTION Dorsum light brown to reddish-brown. Labials: 10–12 upper and 9–11 lower. Nuchal band extends to postocular. Dorsal body has 4–5 dark brown transverse bands bordered by white tubercles, and 20–24 longitudinal rows of dorsal tubercles. Lamellae: 20–25 beneath fourth toe; 30–36 contiguous femoral-precloacal pore-bearing scales in male. Original tail has 12–14 dark brown bands. White caudal bands not immaculate. Hatchling bright yellow with dark brown nuchal and body bands; posterior part of tail white. DISTRIBUTION Occurs in Thailand. HABITS AND HABITAT Nocturnal and arboreal. Female lays two eggs per clutch. Hatchling 38mm SVL; 79mm TL. Inhabits forests, and found on tree trunks, and in crevices, karst boulders and caves.

▪ Typical Geckos ▪

Five-banded Bent-toed Gecko ▪ Cyrtodactylus limajalur SVL 94mm

DESCRIPTION Dorsum greyish-brown overall; 10–12 upper and 9–11 lower labials. Dark brown nuchal loop extends to snout and edged by white line. Nape has narrow, 'V'-shaped band. Five dark bands from nape to posterior body. Three narrow whitish transverse bands on dorsal body. Lamellae: 19–22 beneath fourth toe; 7–8 precloacal and 5–6 femoral scales on each thigh in male. Tail narrow and tapering to a point, with regularly spaced whitish markings. **DISTRIBUTION** Occurs in Malaysia (Borneo). **HABITS AND HABITAT** Nocturnal and arboreal. Usually found on karst boulders in southwestern Sarawak.

Singapore Bent-toed Gecko ▪ Cyrtodactylus majulah SVL 68mm

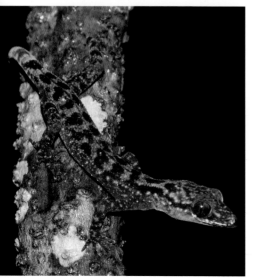

DESCRIPTION Member of the swamp forest-dwelling bent-toed gecko clade. Head brown with yellowish mottling; 8–9 upper and 7–8 lower labials. Dorsum yellowish-brown with oval blotches; 20–23 lamellae beneath fourth toe; 7–11 precloacal pores in male. No enlarged femoral scales or pores. Tail dark brown with cream transverse bands, moderate in size and tapering to a point. Subcaudal dark brown with irregularly shaped beige blotches. Median subcaudal scales not enlarged. **DISTRIBUTION** Occurs in Indonesia and Singapore. **HABITS AND HABITAT** Nocturnal and arboreal. Found on small tree trunks and vegetation about 1.5–2m above the ground. Inhabits lowland rainforests in Singapore and Pulau Bintan, Indonesia.

▪ TYPICAL GECKOS ▪

Marbled Bent-toed Gecko ▪ *Cyrtodactylus marmoratus* SVL 86mm

DESCRIPTION Dorsum light brown with dark brown spots on head and body; 8–13 upper and 7–9 lower labials. Dark temporal streak. Ventral surface brownish; 18–24 lamellae beneath fourth toe. Male has narrow longitudinal precloacal groove; 45–57 femoral precloacal pores. Tail cylindrical, tapering, and covered with small flat scales and a few scattered tubercles; alternating dark and light transverse bands on tail. DISTRIBUTION Occurs in Indonesia and Malaysia. HABITS AND HABITAT Nocturnal and arboreal. Usually seen on tree trunks, and in crevices, rocks and vegetation. Inhabits rainforests from sea level to 1,500m asl.

Mulu Bent-toed Gecko
▪ *Cyrtodactylus muluensis* SVL 88mm

DESCRIPTION Dorsum of this karst-dwelling gecko greyish-brown; 10–13 upper and 8–11 lower labials. Dark brown nuchal loop extends to snout. Dorsal body has 5–8 dark transverse bands and 13–15 longitudinal rows of tubercles. Nuchal loop and body bands without white margins. Ventral uniform light cream; 19–22 lamellae beneath fourth toe; five precloacal pores and no enlarged femoral scales in male. Original tail has nine dark caudal bands separated by narrower grey bands. DISTRIBUTION Occurs in Malaysia (Borneo). HABITS AND HABITAT Nocturnal and arboreal. Known to occur in Mulu National Park. Usually found on karst surfaces and vegetation.

■ TYPICAL GECKOS ■

Oldham's Bent-toed Gecko ■ *Cyrtodactylus oldhami* SVL 77mm

DESCRIPTION Dorsum greyish-brown with dark brown elongated or round spots forming interrupted longitudinal rows on body. Tubercles on dorsal and lateral body bright cream or yellow. Top of head uniform brown. Dark brown nuchal loop extends to postocular and bordered with yellow or white margins; 1–4 preanal pores but no precloacal groove in male. Enlarged femoral scales present, but no pores. Tail dark brown with narrow, white or cream transverse bands. **DISTRIBUTION** Occurs in Myanmar and Thailand. **HABITS AND HABITAT** Nocturnal and arboreal. Usually found near the ground on large rocks and bases of trees. Inhabits lowland forests from sea level to 150m asl.

Butterfly Bent-toed Gecko ■ *Cyrtodactylus papilionoides* SVL 93mm

DESCRIPTION Base dorsum colour light brown with oval and round blotches on top of head; 10–11 upper and 8–10 lower labials. Nuchal band wavy anteriorly. Dorsal body has irregularly shaped dark brown blotches and 12–14 longitudinal rows of tubercles. Paired blotches may be connected medially, forming butterfly-wing pattern. Limbs and digits banded; 4–6 precloacal pores in male; 29–33 contiguous enlarged femoral precloacal scales. Tail has alternating dark brown and whitish transverse bands. **DISTRIBUTION** Occurs in Thailand. **HABITS AND HABITAT** Nocturnal and terrestrial. Usually found on the ground and occasionally on rocks and vegetation to 2m above the ground. Inhabits rocky hills covered with grasses and bushes at about 400m asl.

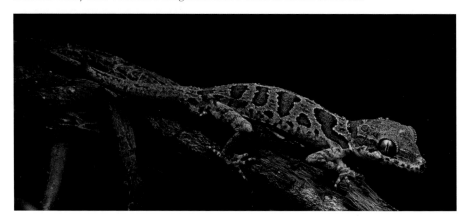

■ TYPICAL GECKOS ■

Philippine Bent-toed Gecko ■ *Cyrtodactylus philippinicus* SVL 98mm

DESCRIPTION Dorsum light yellowish-brown overall, with five irregularly shaped dark transverse bands on body; 6–10 upper and 5–8 lower labials. Dark 'M' marking on neck; 19–26 lamellae beneath fourth toe; 8–12 precloacal pore-bearing scales; 2–5 postcloacal tubercles on each side. Tail tapering, with light and dark transverse bands. **DISTRIBUTION** Occurs in Indonesia (Kalimantan), Malaysia (Borneo) and the Philippines. **HABITS AND HABITAT** Nocturnal and arboreal. Commonly seen on tree trunks close to forest floor. Female lays two eggs per clutch. Hatchlings 30–35mm SVL. Found at sea level to 1,100m asl in riparian habitats near streams.

▪ Typical Geckos ▪

Inger's Bent-toed Gecko ▪ Cyrtodactylus pubisulcus SVL 74mm

DESCRIPTION Dorsum brown to greyish-brown with dark blotches; 10–13 upper and 9–12 lower labials. Dorsal body blotches may be connected to form broad longitudinal lines; 17–22 rows of dorsal tubercles. Limbs and digits have faint blotches and spots; 7–8 precloacal pores in male. No enlarged femoral scales and femoral pores. Original tail pattern alternating dark and cream transverse bands. DISTRIBUTION Occurs in Brunei Darussalam and Malaysia (Borneo). Previously considered a widespread species in Sarawak, a recent study identified two additional lineages: C. hantu (central) and C. miriensis (eastern). HABITS AND HABITAT Nocturnal and arboreal. Usually found on twigs and vegetation; rarely on large tree trunks. Inhabits primary and secondary lowland rainforests in western Sarawak (Borneo) at 30–400m asl.

Malayan Bent-toed Gecko ▪ Cyrtodactylus pulchellus SVL 120mm

DESCRIPTION Overall greyish-brown to reddish-brown. Head uniform brown without spots. Wide dark brown band behind eye connected to nuchal bands bordered with white tubercles; 9–11 upper and 8–10 lower labials. Dorsal body has four transverse dark bands bordered by white tubercles; 22–26 longitudinal rows of dorsal tubercles; 21–26 lamellae beneath fourth toe; 33–39 contiguous enlarged femoral precloacal pores in male. Original tail has 8–10 transverse dark brown bands. Juveniles bright yellowish-orange with dark transverse bands similar to adults. DISTRIBUTION Found in Peninsular Malaysia. HABITS AND HABITAT Nocturnal and arboreal. Often found on tree trunks near streams.

▪ TYPICAL GECKOS ▪

Four-striped Bent-toed Gecko
▪ *Cyrtodactylus quadrivirgatus* SVL 71mm

DESCRIPTION Dorsum grey to pale yellowish-brown. Supraciliary scales yellow and skin above orbits bluish; 10 upper and 10 lower labials. Dark band behind eye extends to lateral line. Dorsal body has four longitudinal lines that may be contiguous or interrupted. Tubercles on head, body, limbs and tail-base. Limbs have dark irregular transverse bands; 0–4 precloacal pores in male. Original tail has alternating dark and light transverse bands. **DISTRIBUTION** Occurs in Indonesia, Malaysia, Singapore and Thailand. **HABITS AND HABITAT** Nocturnal and arboreal. Usually seen on vegetation near the ground. Inhabits rainforests from sea level to 1,400m asl.

Palawan Bent-toed Gecko ▪ *Cyrtodactylus remidiculus* SVL 94mm

DESCRIPTION Dorsum overall dark brown with reticulated pattern on head and nape, and 3–4 narrow, light brown transverse bands on body; 7–11 upper and 6–8 lower labials. Dorsal body has 18–22 rows of paravertebral tubercles. No ventrolateral fold on body. Beneath fourth toe, 19–24 lamellae; 5–8 preanal pores and 8–9 femoral pores in male; 4–6 postcloacal tubercles. Tail dark brown with narrower white transverse bands. **DISTRIBUTION** Occurs in the Philippines (Palawan). **HABITS AND HABITAT** Nocturnal and arboreal. Inhabits riparian habitats of forests at 300–800m asl.

▪ Typical Geckos ▪

Roesler's Bent-toed Gecko ▪ *Cyrtodactylus roesleri* SVL 75mm

DESCRIPTION Overall colour dull brown with small dark spots on head. Dark band with thin yellow margin from snout through eye to nape connected to nuchal band, which is not expanded on neck region; 10–12 upper and 8–10 lower labials. Dorsal

body has 4–5 dark transverse bands bordered by yellow margins. Yellow tubercles on body. Limbs have dark markings; 17–21 lamellae beneath fourth toe; 20–28 contiguous enlarged femoral precloacal pores in male; 5–8 postcloacal spurs in both sexes. Tail has alternating dark and light transverse bands. DISTRIBUTION Occurs in Laos and Vietnam. HABITS AND HABITAT Nocturnal and arboreal. Inhabits limestone forests. Seen on limestone outcrops about 1–2m above the ground.

Sai Yok Bent-toed Gecko ▪ *Cyrtodactylus saiyok* SVL 61mm

DESCRIPTION Dorsum overall light greyish–brown. Skin above eye orbit bluish. Dark brown band behind eye connects to 'V'-shaped nuchal band. Seven upper and 9–10 lower labials. Dorsal body has 3–5 irregularly shaped, dark brown transverse markings that may be contiguous or interrupted medially; 18–19 longitudinal rows of tubercles. Lateral body has faint brown spots; 16–17 lamellae beneath fourth toe; five precloacal pore-bearing scales in male. Tail has seven dark bands. DISTRIBUTION Occurs in western Thailand. HABITS AND HABITAT Nocturnal and arboreal. Inhabits dry evergreen forests at 350–525m asl. Found on small trees, stumps and vegetation.

◾ TYPICAL GECKOS ◾

Sam Roi Yot Bent-toed Gecko ◾ *Cyrtodactylus samroiyot* SVL 67mm

DESCRIPTION Dorsum overall light brown with bands on head and body. Brown stripe from snout through eye to nape, and connected to nuchal band; 10–11 upper and 10 lower labials. Dorsal body-band brown bordered with dark brown and edged with thin yellow margin. Seven contiguous precloacal pore-bearing scales in male. Femoral has row of contiguous, slightly enlarged scales, without pores or pits. Tail has 9–11 dark transverse bands and tapers to pointed tip. DISTRIBUTION Found in Thailand. HABITS AND HABITAT Nocturnal and arboreal. Female lays two hard-shelled eggs per clutch. Inhabits limestone reliefs.

Sanook Bent-toed Gecko ◾ *Cyrtodactylus sanook* SVL 80mm

DESCRIPTION Overall body colour dark brown with irregularly shaped yellowish marking on head and dorsal body; 18–20 longitudinal rows of dorsal tubercles. Wide dark stripe from snout through eye connected to dark nuchal band, bordered with yellow margins. Limbs have small yellowish spots; 19–20 lamellae beneath fourth toe; 32–34 contiguous enlarged femoral precloacal scales. Tapering tail has dark and light transverse bands. DISTRIBUTION Occurs in southern Thailand. HABITS AND HABITAT Nocturnal and arboreal. Inhabits cave entrances to 100m in caves. Usually found on karst cliffs and boulders about 1m above the ground.

▪ Typical Geckos ▪

Tiger Bent-toed Gecko ▪ Cyrtodactylus tigroides SVL 83mm

DESCRIPTION Moderate-sized *Crytodactylus*. Dorsal head yellowish with brown markings. Dorsal body brown with yellowish-cream bands bordered by dark brown margins. Thirteen rows of tubercles at dorsal mid-body; 19–23 lamellae beneath fourth toe; 8–9 precloacal pores in male; 5–7 femoral pores on each thigh in both sexes. Original tail has alternating dark and light transverse bands. DISTRIBUTION Endemic to western Thailand. HABITS AND HABITAT Nocturnal and arboreal. Inhabits lowland limestone hills and usually seen on limestone about 1–1.5m above the ground.

■ TYPICAL GECKOS ■

Tioman Island Bent-toed Gecko
■ *Cyrtodactylus tiomanensis* SVL 83mm

DESCRIPTION Medium-sized *Crytodactylus*. Head yellowish-brown with brown speckling; 8–11 upper and 9–11 lower labials. Pale yellow nuchal band bordered with dark brown. Dorsal body has four pale yellow or light brown transverse bands bordered by irregularly shaped dark brown bands. Lamellae: 20–22 beneath fourth toe. Tail has alternating dark and light transverse bands. **DISTRIBUTION** Endemic to Malaysia. **HABITS AND HABITAT** Nocturnal and arboreal. Inhabits lowland rainforests at 50–150m asl. Found on tree trunks and moss-covered granite boulders.

▪ Typical Geckos ▪

Wayakone's Bent-toed Gecko ▪ *Cyrtodactylus wayakonei* SVL 90mm

DESCRIPTION Dorsum overall dark brown with yellow tubercles on body and limbs. Head has large round or oval blotches bordered by light margins. Nuchal band absent; 7–8 upper and 9–10 lower labials. Dorsal body has yellow reticulated pattern. Tail greyish-black with irregularly shaped transverse beige bands. **DISTRIBUTION** Found in northern Laos and China (Yunnan Province). **HABITS AND HABITAT** Nocturnal and arboreal. Inhabits karst forests at 730–810m asl. Seen in limestone crevices at 1–1.5m above the ground.

Spotted Bent-toed Gecko ▪ *Cyrtodactylus zebraicus* SVL 70mm

DESCRIPTION Overall brown with several oval spots on head. Eleven upper and nine lower labials. Labial scales white with brown markings. Nuchal band extends to behind eye. Dorsal body has dark brown transverse bands bordered by thin yellow margins. Body bands may be interrupted. Eight precloacal pores and two large postcloacal tubercles on each side. Tail alternating black and white; white parts with black specks. **DISTRIBUTION** Occurs in Thailand. **HABITS AND HABITAT** Arboreal and nocturnal. Inhabits lowland rainforests and usually seen on low vegetation and stems.

■ Typical Geckos ■

Sam Roi Yot Leaf-toed Gecko ■ *Dixonius kaweesaki* SVL 42mm

DESCRIPTION Dorsal head and body grey. Pair of black stripes from nostrils run through eyes and continue to upper body to anterior part of tail; light vertebral stripe separates the stripes. Longitudinal rows of tubercles on dorsal (12–13) and ventral (24) body; 10–11 upper and 6–8 lower labials. Digits slender and dilated at tips. Fifteen lamellae beneath fourth toe; 9–11 precloacal pore-bearing scales in male. Tail brownish-orange (light grey if regrown), slender and tapering. **DISTRIBUTION** Occurs in Thailand; endemic to Khao Sam Roi Yot massif, southern Thailand. **HABITS AND HABITAT** Nocturnal and arboreal. Most active 1–4 hours after rainfall. Occurs at 5–300m asl. Usually seen on limestone boulders.

Black-spotted Leaf-toed Gecko
■ *Dixonius melanostictus* SVL 50mm

DESCRIPTION Dorsal body yellowish-brown to lavender-grey, with or without dark spots. Light stripe from snout to forelimb. Dark stripe behind eye to forelimb; 10–11 longitudinal rows of trihedral or keeled tubercles on dorsal body. Wide dark stripe on lower side of body. Nine upper and seven lower labials; 10 lamellae beneath fourth toe; nine precloacal pore-bearing scales in male. Tail slender, tapering and unbanded. **DISTRIBUTION** Endemic to Thailand. **HABITS AND HABITAT** Nocturnal and terrestrial. Hides under rocks and logs by day.

▪ TYPICAL GECKOS ▪

Cha-am Leaf-toed Gecko ▪ *Dixonius pawangkhananti* SVL 42.6mm

DESCRIPTION Relatively slender gecko without ventrolateral folds. Flanks covered with irregular, smooth or slightly conical scales. Dorsal scales similar and distributed among strongly keeled tubercles arranged in 16 or more irregular longitudinal rows. Dorsum greyish base colour with five irregular bars between limb insertions. Black stripe from nostril through eye towards nape, which has two longitudinal blotches followed by one blotch. Tail has grey base colour with two black and 10 orange bars alternating from base towards tip. Supralabials whitish with black spot. Ventral surfaces whitish. DISTRIBUTION Found in Cha-am district, Thailand (around Wat Nikhom Wachiraram). HABITS AND HABITAT Similarly to other *Dixonius* species, found at night in foothills. Uses rock crevices as hideouts. Locally abundant and always found on limestone, never on the ground.

Siamese Leaf-toed Gecko ▪ *Dixonius siamensis* SVL 57mm

DESCRIPTION Dorsal body grey to lavender-brown with large dark spots. Back of head has markings. Supranasals separated by two granular scales; 7–8 upper and 6–7 lower labials, cream-coloured with dark bars; 10–14 longitudinal rows of tubercles on dorsal body; 12–13 lamellae beneath fourth toe; 6–7 precloacal pore-bearing scales in male. Tail may have light and dark bands, narrow transverse dark bands or dark spots. DISTRIBUTION Occurs in Laos, Thailand and Vietnam. HABITS AND HABITAT Nocturnal and terrestrial. Inhabits primary and secondary forests at 0–700m asl.

▪ Typical Geckos ▪

Fehlmann's Four-clawed Gecko ▪ *Gehyra fehlmanni* SVL 51mm

DESCRIPTION Dorsal body light brown with black spots on head, body, limbs and tail. Ventral yellowish-white with small dots; 7–8 upper and lower labials; 22 precloacal femoral pore-bearing scales in male. Tail slightly shorter than SVL, tapering and narrow. Ventral tail has single row of enlarged subcaudal scales and heavily pigmented brown. Juveniles have pairs of dark spots and light, round spots on dorsal body; tail banded. **DISTRIBUTION** Occurs in Thailand and Vietnam. **HABITS AND HABITAT** Nocturnal, and both arboreal and terrestrial.

Kanchanaburi Four-clawed Gecko ▪ *Gehyra lacerata* SVL 55mm

DESCRIPTION Dorsal body grey with dark spots and two rows of light round spots; dorsal scales granular and without tubercles. Twelve upper labials and 10–11 lower labials; 18–22 precloacal femoral pore-bearing scales in male. Ventral body cream without markings. Tail distinctly shorter than SVL. **DISTRIBUTION** Occurs in Thailand and Vietnam. **HABITS AND HABITAT** Nocturnal, and possibly both arboreal and terrestrial. Usually found hiding under rocks and logs.

▪ Typical Geckos ▪

Common Four-clawed Gecko ▪ *Gehyra mutilata* SVL 60mm

DESCRIPTION Medium-sized *Gehyra* species with moderately depressed head and body. Head width as broad as body. Dorsal body colour light grey or brown, with or without dark spots. Skin fold on posterior margin of thigh. Digits webbed at bases and broadly dilated. All digits have claws; first (inner) claw concealed. Male has 32–40 precloacal femoral pore-bearing scales. Juveniles have very small black and white spots. DISTRIBUTION Found in Cambodia, Indonesia, Malaysia, the Philippines, Thailand, Timor-Leste and Vietnam. HABITS AND HABITAT Human commensal gecko. Mainly feeds on insects and arachnids, but also on flower nectar. Female usually lays two eggs, which are adhered to wall crevices. Occurs from sea level to 600m asl. NOTE The skin of this gecko is easily torn if handled improperly.

Golden Gecko ▪ *Gekko badenii* SVL 108mm

DESCRIPTION Dorsal body dark olive-grey to yellowish-brown, with 6–8 light narrow, interrupted transverse bands; 11–14 upper and 11–12 lower labials; 14–19 lamellae beneath fourth toe; 10–18 precloacal pore-bearing scales in male. Tail uniform in colour, with or without light bands. DISTRIBUTION Found in Vietnam. HABITS AND HABITAT Occurs from sea level to 1,000m asl in forests and inland rock outcrops.

◾ Typical Geckos ◾

Brown's Forest Gecko ◾ *Gekko browni* SVL 66mm

DESCRIPTION Small gecko with slender, elongated body that is greyish-brown with irregularly shaped dark markings on back. Discontinued light orange ventral line may be present; 13–15 upper and 13–14 lower labials. Limbs have expanded skin folds. Digits strongly webbed; 16–19 scansors beneath fourth toe; 28–32 precloacal femoral pore-bearing scales. Tail depressed with recurved scales on both sides. DISTRIBUTION Occurs in Peninsular Malaysia; recently also discovered in southern Thailand. HABITS AND HABITAT Arboreal, nocturnal and forest obligate. Occupies higher stratum of forests, thus seldom observed. Usually found on large tree trunks about 2–3m above the ground.

Luzon Karst Gecko ◾ *Gekko carusadensis* SVL 97mm

DESCRIPTION Moderate-sized gecko. Dorsal body grey with no dark mottling or few transverse bars; 16–18 tubercles per row at mid-body. Can be distinguished from the Mindoro Narrow-disked Gekko (p. 89), the most morphologically similar species by: male has 46–50 precloacal femoral pores arranged in uninterrupted series (v 52–66), and 18–20 scansors beneath fourth toe (v 12–14). Tail has transverse bars, darkest posteriorly. DISTRIBUTION Endemic to the Philippines (Luzon). HABITS AND HABITAT Nocturnal. Inhabits karst outcrops and caves at low elevation. Recently described; very limited information available on its natural history.

Typical Geckos

Tokay Gecko ◾ *Gekko gecko* SVL 166mm

DESCRIPTION Large gecko with grey base colour on head, body, legs and tail. Head large and distinct from neck. Orange spots on head, body and limbs, fading on tail, often combined with eight transversely alligned rows of white spots. White spots often absent on head. Tail as long as SVL. Juveniles generally more bold in colour than adults and have transverse white bands on body and tail. Colour can vary across geographic range, with some animals displaying very dark to almost black base colour with almost no orange spots present. Some individuals do not have any spots or have merely very faint spots. Colour and pattern can vary throughout the day or according to surroundings: more substrate-matching colour in natural habitats, compared to when living in human settlements. **DISTRIBUTION** Occurs in Cambodia, Indonesia, Laos, Malaysia, Myanmar, the Philippines, Thailand, Timor-Leste and Vietnam. **HABITS AND HABITAT** Nocturnal and arboreal. Commonly seen in rural houses on walls and in secondary forests on tree trunks. Has been observed in both lowland and hill dipterocarp forests. Can be found hiding in rock cracks or under tree bark by day. Sometimes found up to 5m above the ground at night. Call, *to-kay*, is repeated several times and is diagnostic for the species. This barking behaviour is often considered a nuisance by humans. Mainly feeds on invertebrates (cockroaches, beetles, crickets), and opportunistically on vertebrates (geckos, small birds, house mice). Female lays two eggs that are adhered inside tree holes. Hatchling size about 40mm SVL. **NOTE** This species is harvested throughout its range in large quantities to supply the international trade for pets (live) and traditional medicine (dried), resulting in a listing on CITES Appendix II regulating its international trade.

■ TYPICAL GECKOS ■

Horsfield's Parachute Gecko ■ *Gekko horsfieldii* SVL 74mm

DESCRIPTION Parachute geckos, formerly genus *Ptychozoon*, have broad skin-flaps on neck, sides of body and limbs. Digits extensively webbed and both sides of tail have rounded lobes. This species is light greyish-brown with dark markings on dorsal body; 11–13 lamellae beneath fourth toe; 10–11 enlarged pore-bearing scales in male; 8–11 femoral scales. Caudal lobes strongly pointing backwards. Tail has 4–5 dark transverse bands. Terminal tail-flap not expanded. **DISTRIBUTION** Found in Brunei Darussalam, Indonesia, Malaysia, Myanmar, Singapore and Thailand. **HABITS AND HABITAT** Nocturnal and arboreal. Occurs in lowland primary and secondary forests (below 300m asl), but seen in forest edges near human dwellings.

Kaeng Krachan Parachute Gecko
■ *Gekko kaengkrachanense* SVL 86mm

DESCRIPTION Dorsal body has three dark chevron markings and no enlarged tubercles; 15–17 lamellae beneath fourth toe; 14–19 enlarged pore-bearing scales in male. Tail tapering and without long terminal flap. **DISTRIBUTION** Endemic to western Thailand. **HABITS AND HABITAT** Arboreal and nocturnal. Occurs in montane forests of Thailand. Commonly seen on walls of man-made structures hunting insects attracted by artificial lights.

▪ Typical Geckos ▪

Kuhl's Parachute Gecko ▪ *Gekko kuhli* SVL 108mm

DESCRIPTION Largest known parachute gecko. Body has two rows of ovoid tubercles; 12–16 lamellae beneath fourth toe; 14–32 enlarged pore-bearing scales in male. Terminal tail-flap elongated (2.1–3.0cm) and widely expanded. DISTRIBUTION Occurs in Indonesia, Malaysia, Myanmar, Singapore and Thailand. HABITS AND HABITAT Seen in primary forests, usually near streams, as well as disturbed forests.

Burmese Parachute Gecko
▪ *Gekko lionotum* SVL 96mm

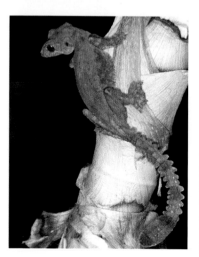

DESCRIPTION Colour variable, from olive-grey to brown. Dorsal body has four dark bands bordered with thin black wavy lines ('M' shaped) posteriorly, and irregularly shaped light vertebral markings. No tubercles on body and tail; 16–17 lamellae beneath fourth toe; 20–22 enlarged pore-bearing scales in male; 18–23 tail-lobes pointing backwards, and minimal size reduction posteriorly. DISTRIBUTION Found in Laos and Myanmar. HABITS AND HABITAT Forest obligate, nocturnal and arboreal. Seen on tree trunks of various sizes 1.5–2.0m above the ground.

▪ TYPICAL GECKOS ▪

Mindoro Narrow-disked Gecko
▪ *Gekko mindorensis* SVL 88mm

DESCRIPTION Moderate-sized gecko with slightly depressed body. Dorsal body grey with thin dark transverse bands. Tubercles at mid-body: 16–20 per row. Male has 52–66 precloacal femoral pores arranged in uninterrupted series, and 12–14 scansors beneath fourth toe. Adult female slightly smaller than male. DISTRIBUTION Found in the Philippines. HABITS AND HABITAT Occurs from sea level to 900m asl in and near entrances of limestone caves. Eggs usually glued on cave walls.

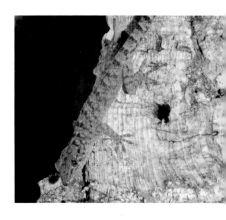

Spotted House Gecko ▪ *Gekko monarchus* SVL 100mm

DESCRIPTION Overall brown to grey with dark spots on head and body. Back of head has 'W' marking. Dorsal body has series of parallel spots. Tubercles on body yellow or white. Male has 31–40 precloacal femoral pore-bearing scales; 13–15 scansors beneath fourth toe. Tail has white transverse bands distally. DISTRIBUTION Found in Brunei Darussalam, Indonesia, Malaysia, the Philippines, Singapore and Thailand. HABITS AND HABITAT Nocturnal and arboreal. Human commensal species commonly seen in houses, buildings and gardens. Feeds primarily on insects. Female adheres two eggs per clutch on walls, rock crevices and logs. Incubation may take to 120 days.

▪ Typical Geckos ▪

Nutaphand's Gecko ▪ *Gekko nutaphandi* SVL 116mm

DESCRIPTION Large gecko with elongated, robust body. Overall greyish-brown. Dorsal body has bright white spots arranged in eight transverse bands and 14 rows of enlarged tubercles. Iris deep brick-red; 12–14 precloacal pore-bearing scales. Tail has alternating dark and light transverse bands, with dark bands at least two times broader. DISTRIBUTION Found in Thailand. HABITS AND HABITAT Inhabits limestone and surrounding bamboo forests. Usually seen on walls of caves and on bamboos.

Palawan Gecko
▪ *Gekko palawanensis* SVL 66mm

DESCRIPTION Small gecko with grey or brown dorsal body and paired dark spots. Scales on head small and uniform in size; 64–70 precloacal femoral pore-bearing scales. No webbing between digits; 16–19 scansors beneath fourth toe. DISTRIBUTION Found in the Philippines (Palawan). HABITS AND HABITAT Arboreal and nocturnal. Occurs from sea level to 900m asl. Usually seen close to the ground on tree trunks, beneath loose bark and on roofs of small caves.

▪ Typical Geckos ▪

Palmated Gecko
▪ *Gekko palmatus* SVL 83mm

DESCRIPTION Robust body and broad head. Dark round or oval spots on occipital and nuchal region. Dorsal body greyish-brown with dark mid-dorsal spots (pair of oval spots on some individuals), followed by cream round spots; 11–15 upper labials and 9–13 lower labials. Tubercles on dorsal body, but absent on limbs. Sides of body and limbs have small light spots. Ventral yellowish-cream with dark dots; 11–17 lamellae beneath fourth toe; 24–32 precloacal femoral pore-bearing scales. Tail has alternating light and dark transverse bands. DISTRIBUTION Occurs in northern and central Vietnam. HABITS AND HABITAT Nocturnal and arboreal. Occasionally seen on walls or inside houses.

Sandstone Gecko ▪ *Gekko petricolus* SVL 100mm

DESCRIPTION Dorsal body yellowish with longitudinal rows of round white spots. Head greyish-lavender; 12–15 upper and 10–12 lower labials; 9–11 precloacal pore-bearing scales in male. Tail depressed and slender, with or without dark spots. DISTRIBUTION Found in Cambodia, Laos and Thailand. HABITS AND HABITAT Occurs in sandstone hills. Hides between horizontal crevices of rocks, usually in upside-down position. Female lays two hard-shelled eggs and adheres them to rocks.

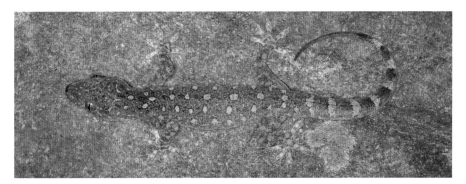

▪ TYPICAL GECKOS ▪

Sabah Parachute Gecko ▪ *Gekko rhacophorus* SVL 74mm

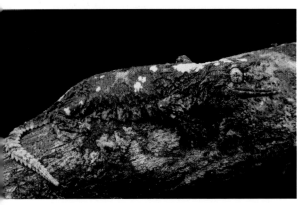

DESCRIPTION Lateral skinfolds on head absent. Digits half webbed. Irregular spinose tubercles on body; 11–13 lamellae beneath fourth toe; 12–18 enlarged pore-bearing scales in male. Tail sharply tapered with jagged edge tail-lobes. **DISTRIBUTION** Endemic to Borneo (Malaysia). **HABITS AND HABITAT** Arboreal, nocturnal and forest obligate. Inhabits submontane and montane forests at 640–1,400m asl.

Phong Nha-Ke Bang Gecko ▪ *Gekko scientiadventura* SVL 73mm

DESCRIPTION Small-bodied gecko distinguished from all other *Gekko* species in Vietnam by lack of tubercles on head, dorsal body and tail. Dorsal body brown with seven light oval spots. Throat marbled with yellowish reticulation; 14–17 lamellae beneath fourth toe; 5–8 precloacal pore-bearing scales in male. Tail slightly depressed, with 7–10 light transverse bands. **DISTRIBUTION** Currently known only from Vietnam. **HABITS AND HABITAT** Nocturnal, arboreal and forest obligate. Female lays two eggs in rock crevices in communal egg-laying sites. Usually seen on vegetation (on rocks for the sympatric **Phongnhakebang Bent-toed Gecko** *Cyrtodactylus phongnhakebangensis*) 1–2.5m above the ground in primary and bordering forests over karst limestone.

▪ Typical Geckos ▪

Smith's Green-eyed Gecko ▪ *Gekko smithii* SVL 190mm

DESCRIPTION Very large gecko with large head, robust body, and green to olive eyes. Overall dorsal colour light greyish-brown to olive-brown, with series of transverse white spots or dark transverse bands. Limbs and tail may be darker, with irregular light spots. Hatchlings dark with round white spots on body and white transverse bands on tail. DISTRIBUTION Found in Brunei Darussalam, Indonesia, Malaysia, Myanmar, Singapore and Thailand. HABITS AND HABITAT Nocturnal and insectivorous. Female lays two spherical eggs about 2.3cm in diameter, which are adhered to tree trunks. Hatchling 7.8cm TL. Occurs in lowland and mid-elevation rainforests, but occasionally seen near human settlements.

Three-banded Parachute Gecko ▪ *Gekko trinotaterra* SVL 71mm

DESCRIPTION Head and body depressed. Dorsal body colour light grey mottled with dark brown blotches and tiny black spots. Can be distinguished from other parachute geckos by three dark, transverse ('M'-shaped) wavy bands on back; 19–21 precloacal femoral pore-bearing scales arranged in contiguous series. Tail flattened, with 15–16 lobes on each side. Limbs and tail have dark bands. Tail-tip enlarged. DISTRIBUTION Found in Cambodia, Thailand and Vietnam. HABITS AND HABITAT Nocturnal and arboreal. Inhabits primary and secondary rainforests. Has been seen on large tree trunks at 2–3m above the ground, actively hunting winged termites after rainfall in Thailand.

▪ Typical Geckos ▪

Lined Gecko ▪ *Gekko vittatus* SVL 110mm

DESCRIPTION Large gecko with slender body and characteristic 'Y'-shaped white marking when viewed from above. Overall light brown. Small, flat tubercles on head, body, limbs, throat and lateral skin-folds. Tail has transverse dark and light bands; dark bands twice broader than light ones; 12–26 lamellae beneath fourth toe; 39–65 precloacal femoral pore-bearing scales in male. **DISTRIBUTION** Occurs in Indonesia. **HABITS AND HABITAT** Arboreal and nocturnal. Female lays two eggs per clutch. Usually seen on tree branches 2m above the ground.

Brook's House Gecko ▪ *Hemidactylus brookii* SVL 56mm

DESCRIPTION Taxonomic status unclear and heavily debated, might include previously synonymized *H. murrayi*. Body light to dark brown, with or without dark spots. Ear opening large, oval and slanted; 8–10 upper and 9–11 lower labials. Dorsal body has large tubercles forming 16–19 longitudinal rows. Two series of 12–13 precloacal femoral pores separated in middle by non-pore-bearing scale in male; 7–8 lamellae beneath fourth toe. Anterior portion of tail has recurved conical spines. Subcaudal scales longer than high and cover entire ventral surface of tail. **DISTRIBUTION** Occurs in Indonesia, Malaysia, Myanmar, the Philippines and Thailand. **HABITS AND HABITAT** Arboreal, nocturnal and human commensal. Common in houses and other man-made structures. Usually seen at night on walls while hunting for insects attracted by artificial lights.

■ Typical Geckos ■

Mocquard's House Gecko ■ *Hemidactylus craspedotus* SVL 55mm

DESCRIPTION Light mottled brown with discontinued dark dorsolateral lines and irregularly shaped light markings. Dark stripe runs through eye. Sides of neck, lateral body and tail have small skin-flap. Ventral body bright yellow. Tail depressed and tapering. Subcaudal scales reddish-orange. DISTRIBUTION Found in Indonesia, Malaysia, Singapore and Thailand. HABITS AND HABITAT Arboreal and nocturnal. Inhabits rainforests and swamp forests. Usually seen on tree trunks as well as near and in human dwellings.

Common House Gecko ■ *Hemidactylus frenatus* SVL 60mm

DESCRIPTION Slender gecko with depressed body and head slightly distinct from neck. Dorsal body colour variable, from grey to light brown, with or without dark blotches or scattered small spots. Digits moderately dilated, with claws, and not webbed; 9–11 divided scansors beneath fourth toe. Tail slightly longer than SVL. Sides of tail have small spiny scales at intervals of 7–12 scales; regenerated tail does not have spiny scales. DISTRIBUTION Found in Brunei Darussalam, Cambodia, Indonesia, Laos, Malaysia, Myanmar, the Philippines, Singapore, Thailand and Vietnam. HABITS AND HABITAT Arboreal, nocturnal and human commensal. Very common in human settlements, even in urban areas. Often congregates near light fixtures to hunt insects attracted to light. Female lays pair of round, hard-shelled eggs. Incubation period 50–70 days. Hatchling 19–22mm SVL; 37–42mm TL.

■ TYPICAL GECKOS ■

Indopacific House Gecko ■ *Hemidactylus garnotii* SVL 66mm

DESCRIPTION All-female, parthenogenetic gecko (juvenile shown). Head and body depressed. Dorsal body greyish-brown with dark or faint dorsolateral stripes from neck to tail-base. Light blotches on lines may be present; 12–14 scansors beneath fourth toe. Tail moderately depressed.

DISTRIBUTION Found in Indonesia, Laos, Malaysia, Myanmar, the Philippines, Thailand and Vietnam. HABITS AND HABITAT Arboreal and nocturnal. Adaptable to disturbed habitats and can be seen on rock outcrops and vegetation near human settlements in rural areas.

Flat-tailed House Gecko ■ *Hemidactylus platyurus* SVL 64mm

DESCRIPTION Moderate-sized, with depressed body and tail. Head tapering and slightly distinct from neck; 9–12 upper labial scales, with 8th–10th beneath centre of eye. Body light grey or brown with spots and blotches. Prominent skin-folds on sides of body and posterior edges of hindlimbs; 7–9 lamellae beneath fourth toe. Tail very depressed, with marginal fringes. DISTRIBUTION Found in Brunei Darussalam, Cambodia, Indonesia, Laos, Malaysia, Myanmar, the Philippines, Singapore, Thailand and Timor-Leste. HABITS AND HABITAT Mainly arboreal and nocturnal. Human commensal and common in urban and rural areas; hunts insects near artificial lights. Female lays pair of round, hard-shelled eggs. Hatchling 20–25mm SVL.

■ Typical Geckos ■

Roti Island House Gecko ■ *Hemidactylus tenkatei* SVL 62mm

DESCRIPTION Overall body colour light brown with small dark spots; 9–12 upper and 9–10 lower labials. Dorsal body has 16–20 longitudinal rows of tubercles, with largest tubercle 11–13 times larger than surrounding granules; 7–9 lamellae beneath fourth toe; 5–8 precloacal femoral pores on each thigh, separated in middle by 5–7 non-pore-bearing scales. Tail 10–25 per cent longer than SVL, depressed and tapering. Anterior portion of tail has spiny scales on both sides; subcaudal scales significantly narrower than in Brook's House Gecko (p. 94) and surrounded by smaller scales. DISTRIBUTION Occurs in Indonesia, Myanmar and Timor-Leste. HABITS AND HABITAT Arboreal, nocturnal and human commensal. Commonly seen on walls and rock outcrops, and in houses.

Sumatran Slender Gecko
■ *Hemiphyllodactylus margarethae* SVL 47mm

DESCRIPTION Body robust and slightly compressed. Head moderately large. Lamellae: 4–7 beneath first toe; lamellar formula of digits II–V on hand (4-4-4-4) and foot (4-5-5-5). Adult female often has 0–12 precloacal pores; precloacal pores (11) and femoral pores (0–29) in male never in contact; 1–2 cloacal spurs. Tail round in cross-section. DISTRIBUTION Found in Indonesia. HABITS AND HABITAT Arboreal, nocturnal and forest obligate.

■ TYPICAL GECKOS ■

Titiwangsa Slender Gecko
■ *Hemiphyllodactylus titiwangsaensis* SVL 62mm

DESCRIPTION Body robust and slightly compressed. Dorsal body light greyish-brown to medium brown, with irregularly shaped dark brown bars. Lamellae: 5–8 beneath

first toe; lamellar formula of digits II–V on hand (3-4-4-4) and foot (4-5-5-5). Precloacal and femoral pore-bearing scales (17–39) contiguous; 1–4 cloacal spurs. Tail round to elliptical in cross-section, usually lighter coloured than body, and with light and dark bands. DISTRIBUTION Found in Malaysia (Pahang). HABITS AND HABITAT Arboreal, nocturnal and forest obligate. Occurs in montane forests in Pahang, Malaysia.

Indopacific Slender Gecko ■ *Hemiphyllodactylus typus* SVL 46mm

DESCRIPTION All-female, parthenogenetic species. Body slender and elongated. Dorsal body greyish-brown to reddish-brown, with series of dark brown transverse bars or blotches. Lamellae: 5–6 beneath first toe; lamellar formula of digits II–V on hand (3-4-4-4) and foot (4-4-5-4);

1–5 rounded cloacal spurs. DISTRIBUTION Occurs in Brunei Darussalam, Indonesia, Malaysia, Myanmar, the Philippines, Singapore and Thailand. HABITS AND HABITAT Arboreal, nocturnal and forest obligate. Mainly seen on tree trunks. When disturbed, retreats to tree crevices or loose bark.

▪ Typical Geckos ▪

Common Smooth-scaled Gecko
▪ *Lepidodactylus lugubris* SVL 45mm

DESCRIPTION All-female, parthenogenetic species. Males rarely observed, but are infertile. Dorsal body light brown to greyish-brown, with small dark spots and wavy lines. Pair of dark, round or oval spots behind neck. Beneath fourth toe, 12–18 scansors. Tail has dark, narrow, wavy transverse bands, usually lighter in colour than body. **DISTRIBUTION** Occurs in Indonesia, Malaysia, Myanmar, the Philippines, Singapore, Thailand and Vietnam. **HABITS AND HABITAT** Arboreal, inhabiting mainly coastal habitats such as mangroves and beach forests. Usually seen on trees and bare rocks. Female lays two eggs that are adhered between leaf axils. Hatchling 18mm SVL.

Sabah Scaly-toed Gecko ▪ *Lepidodactylus ranauensis* SVL 45mm

DESCRIPTION Body greyish-brown with light, round vertebral spots and a few small dark spots. Nine upper labials and 9–10 lower labials. Fifteen scansors beneath fourth toe. Tail subcylindrical with pair of dark spots at base. Thirty-seven precloacal femoral pore-bearing scales and pair of cloacal spurs in male. **DISTRIBUTION** Found in Malaysia (Borneo). **HABITS AND HABITAT** Nocturnal and arboreal. Occasionally seen on walls of man-made structures.

▪ Typical Geckos ▪

Southern Philippine False Gecko
▪ *Pseudogekko pungkaypinit* SVL 77mm

DESCRIPTION Large false gecko with overall greyish-brown colour and series of light brown stripes on lateral body; 16–20 upper and 17–19 lower labial scales; 17–21 scansors beneath fourth toe; 17–20 precloacal pores in adult male. **DISTRIBUTION** Found in the Philippines (Bohol, Leyte and Mindanao). **HABITS AND HABITAT** Arboreal forest obligate. Usually seen on shrubs 2–4m above the ground.

Polillo False Gecko ▪ *Pseudogekko smaragdinus* SVL 64mm
DESCRIPTION Slender, brightly coloured gecko. Body colour bright yellow to yellow-orange (undisturbed) to yellow-green (disturbed). Head and body have round and oval dark spots and a few scattered smaller white spots; 16–22 scansors beneath fourth toe; 32–41 enlarged precloacal pores in adult male. Tail brownish-orange with white bands. **DISTRIBUTION** Occurs in the Philippines (Luzon and Polillo). **HABITS AND HABITAT** Arboreal forest obligate. Usually found between leaf axils of *Pandanus* trees.

▪ EURASIAN LIZARDS ▪

LACERTIDAE (LACERTINAE) (EURASIAN LIZARDS)
The Eurasian lizards of the subfamily Lacertinae include the east Asian and Southeast Asian genus *Takydromus* (oriental grass lizards). *Takydromus* species are characterized by their very long tails and keeled scales.

Asian Grass Lizard ▪ *Takydromus sexlineatus* SVL 65mm
DESCRIPTION Mostly characterized by slender body in combination with tail that can be 3–5 times longer than body. Brownish vertebral stripe on dorsum, and yellowish stripe from orbit to flanks. Paravertebral stripe extends beyond tail-base. Lateral sides can be greenish-yellowish with pale brown sheen. **DISTRIBUTION** Common in Cambodia, Indonesia (Sumatra, Natuna archipelago, Kalimantan, Java and Bali), Laos, Malaysia, Myanmar, Thailand and Vietnam. **HABITS AND HABITAT** Diurnal and arboreal. Mainly found in open areas such as grassland, and marshes up to mid-hill level. Generally occurs from sea level to 850m asl.

■ Borneo Earless Monitor ■

Lanthanotidae (Borneo Earless Monitor)
Only one species is included in the Lanthanotidae, a secretive and semi-fossorial and semi-aquatic lizard found in lowland rainforests on the island of Borneo.

Borneo Earless Monitor ■ *Lanthanotus borneensis* SVL 200mm
DESCRIPTION Slender, elongated monitor lizard with brownish base colour and dark vertebral stripe. Ventral side lighter in colour. Eyes reduced and iris is bluish. Six parallel rows of enlarged scales that run from just behind eyes to tail-tip. Lateral sides also have enlarged scales. **DISTRIBUTION** Endemic to Borneo, Malaysia (Sarawak), Indonesia (Kalimantan); probably also in Brunei Darussalam. **HABITS AND HABITAT** Secretive monitor lizard that is strictly nocturnal and hides by day in deep burrows along river banks or under rocks or logs. Mainly found near lowland streams and marshes, but can also occur near streams on agricultural land.

■ Flat-footed Lizards ■

Pygopodidae (Flat-footed Lizards)

Often mistaken for snakes, these lizards have vestigial hindlimbs reduced to scaly flaps, but completely lack forelimbs. Moving in a snake-like fashion, they prey on other lizards, and on arthropods. Although most pygopodids are found in Australia, two species occur on the island of New Guinea, of which one is endemic to the island.

Jicar's Snake Lizard ■ *Lialis jicari* SVL 310mm

DESCRIPTION Snake-like lizard with distinctive flat-tipped, pointed snout that is turned slightly upwards. Colour varies, and can be greyish, yellowish or pinkish-brown on dorsum, speckled with black. Faint darker vertebral stripe that divides into two on head. Well-defined dorsolateral stripe from snout-tip, through eye, slightly fading towards back. **DISTRIBUTION** Found in Indonesia (and Papua New Guinea), in eastern and northern part of Papua Province. **HABITS AND HABITAT** Largely diurnal. Relatively common throughout its range and found in grassland, savannah woodland, swamp areas and edges of lowland rainforests. Also occurs in gardens. Found mainly in coastal provinces, but also in some highland areas, to 1,600m asl.

SKINKS

SCINCIDAE (SKINKS)

This is one of the largest lizard families, and is often divided into several subfamilies. Southeast Asia is home to the burrowing skinks (Scincinae), social skinks (Egerninae), Australasian skinks (Sphenomorphinae), sun and ground skinks (Mabuyinae), Austral and snake-eyed skinks (Eugongylinae), and Afro-Asian supple, writhing and tree skinks (Lygosominae). Many of these subfamilies are distributed widely across the globe, inhabiting a wide range of habitats.

Grey Tree Skink
- *Dasia grisea* SVL 130mm

DESCRIPTION Relatively slender skink with elongated snout. Light or dark brown, with 8–14 narrow dark rings on dorsum. Tail dark banded. Juveniles have a more distinctive dark pattern than adults. Ventral bright green. **DISTRIBUTION** Found in Peninsular Malaysia and Sarawak, Indonesia (Sumatra and Kalimantan), Singapore and the Philippines (Mindoro, Marinduque, Semirara and Luzon). Recently also discovered in southern Thailand. **HABITS AND HABITAT** Diurnal and arboreal. Mainly found 2–5m up tree trunks. Inhabits lowland dipterocarp forests.

▪ SKINKS ▪

Olive Tree Skink
▪ *Dasia olivacea* SVL 115mm

DESCRIPTION Robust tree skink with yellowish-olive or greenish-brown dorsum. May have black-spotted pattern; ventral area creamish or greenish, without pattern. Juveniles golden-yellowish, with 13–16 dark bands of three scales wide. DISTRIBUTION Widespread in Cambodia, Indonesia (for example, Sumatra, Java and Bali), Laos, Myanmar, Peninsular Malaysia, Singapore, the Philippines (Mindoro) and Vietnam. HABITS AND HABITAT Diurnal and arboreal. Mainly found at forest edges near clearings, where it uses loose bark for shelter. Inhabits open forests to 1,200m asl.

Striped Tree Skink
▪ *Dasia vittata* SVL 100mm

DESCRIPTION Relatively large and robust skink primarily identified by black-and-white striped head and neck. Small yellowish stripe from tip of the snout, between the eyes, stops just behind head. Two stripes start at tip of snout, run above eye and fade towards back. Remaining two stripes on either flank start at tip of snout and fade near forelimbs. Rest of body an olive-brown colour with pale flecks. DISTRIBUTION Found on Borneo, on coasts of Peninsula Malaysia, Indonesia and Brunei Darussalam. HABITS AND HABITAT Diurnal and arboreal. Can be found in both primary and secondary forests, but also in coastal areas. Forages around tree trunks for insects, but otherwise rarely seen near the ground.

▪ Skinks ▪

Mangrove Skink
▪ *Emoia atrocostata* SVL 97.5mm

DESCRIPTION Elongated skink with greyish-olive dorsum that can be flecked with dark brownish or greyish colour. Ventral area bluish, greyish or cream, with dark markings on throat. DISTRIBUTION Found in Brunei Darussalam, Indonesia (Kalimantan, Java, Sumatra, Mentawai and Natuna archipelagos), Malaysia, Singapore and Vietnam. Also Papua New Guinea, Solomon Islands and Australia. HABITS AND HABITAT Diurnal species strongly associated with coastal regions, sandy or rocky beaches and mangrove forests. Can be seen basking on trees by day. Uses hollow tree trunks for shelter during high tide.

Pacific Blue-tailed Skink
▪ *Emoia caeruleocauda* SVL 65mm

DESCRIPTION Small, robust-looking skink with dark base colour. Very bright, light blue tail and three golden stripes that radiate from just above eye (two stripes) or tip of snout (one), over dorsum towards tail, slowly changing colour to blue. Fading black, stripe-like markings on tail. DISTRIBUTION Widespread in most of the Pacific. In Southeast Asia, found in Indonesia (Kalimantan, Sulawesi, Maluku and Papua), Malaysia (Sabah), and the Philippines (Camiaran, Balabac, Tulian and Palmas). HABITS AND HABITAT Diurnal and terrestrial. Occurs in wide range of habitats, including forest clearings, sandy beaches and rural gardens. In the Philippines seems to be associated with small rocky islands.

▪ Skinks ▪

Copper-tailed Skink ▪ *Emoia cyanura* SVL 61mm

DESCRIPTION Typical *Emoia* species, with slender body and short snout. Dorsum brownish-black or can be dark brown. Three light dorsolateral stripes from snout-tip towards tail, fading to blue or brown. Tail greenish, bluish or copper. DISTRIBUTION Occurs throughout the Pacific. In Southeast Asia, found in Indonesia (Kalimantan, Sulawesi, Seram, Sula archipelago and Halmahera). HABITS AND HABITAT Diurnal and terrestrial. Often found on or near leaf litter, in remnants or remains of palm-tree vegetation. Can be seen climbing low vegetation. Inhabits wide range of habitats, including coastal scrub, gardens, plantations and other secondary vegetation.

Red-tailed Swamp Skink ▪ *Emoia ruficauda* SVL 53.6mm

DESCRIPTION Part of the *Emoia cyanura* species group (that is, with nasal bones not fused) and *caeruleocauda* subgroup (rounded or moderately thinned subdigital lamellae). Easily identified by thinned subdigital lamellae (55–63) and colour pattern. Five yellow lines visible from tip of snout to base of tail. Dorsum body black, tail and limbs bright red/orange. DISTRIBUTION Endemic to the Philippines, where it occurs on Mindanao. HABITS AND HABITAT Little is known about this diurnal species. Can be relatively common in areas where it occurs. Previously found in disturbed and undisturbed primary lowland forests (200–300m asl). Also found in tall grass near lakes or rivers, on broad leaves on low shrubs and on forest floor.

▪ SKINKS ▪

Antoni Night Skink ▪ *Eremiascincus antoniorum* SVL 67mm

DESCRIPTION Slender skink with brown base colour and indistinct dark longitudinal lines on dorsum. Flanks have black and white specks, with black spots closest to

dorsolateral line. Posterior from eye towards back has large black markings that reduce in size behind forelimbs. Lips have dark markings, creating barred effect. Ventral surface yellowish. DISTRIBUTION Occurs in Indonesia on the island of Timor, and possibly also in Timor-Leste. HABITS AND HABITAT Crepuscular. Found under stones in damp places.

Bar-lipped Sheen Skink ▪ *Eugongylus rufescens* SVL 143mm

DESCRIPTION Robust-looking, smooth-scaled skink with tail almost as thick as body. Dorsum brown with either no pattern or thin, regular transverse banding. Dark bars visible on supralabials and infralabials. 'V'-shaped mark on gular region. Juveniles have numerous transverse cream bars. DISTRIBUTION Found in Indonesia (Sulawesi, Seram, Ambon and Papua, including West Papua). Also Papua New Guinea, Solomon Islands and Australia. HABITS AND HABITAT Crepuscular or diurnal. Inhabits primary forests between lowlands and submontane limits, but also found in or around coconut plantations.

▪ SKINKS ▪

Long-tailed Sun Skink ▪ *Eutropis longicaudata* SVL 140mm

DESCRIPTION Robust-looking skink with brownish dorsum. Dark band on flanks from eye towards tail, with random pale spots. Pale creamish – sometimes with random small darker spots – under dark stripe and venter. Distinguished from the Common Sun Skink (p. 111) by tail that is at least twice the body length. Juveniles similar in colour to adults, with the exception of the broad lateral stripe, which is black in juveniles and lacking small pale spots laterally. **DISTRIBUTION** Found in Cambodia, Laos, Thailand, Peninsular Malaysia and Vietnam. Also China. **HABITS AND HABITAT** Diurnal. Most easily seen basking on rocks or fallen trees in late morning, or after rain showers. Inhabits lowland areas in heavily disturbed habitats like agricultural fields, parks, roadsides and grassland. Can easily be seen foraging among surface debris. Usually found near human settlements below 500m asl.

▪ Skinks ▪

Bronze Skink ▪ *Eutropis macularia* SVL 75mm

DESCRIPTION Relatively slender skink with distinct head. Dorsum has bronze-brown base colour and can be spotted. Flanks darker, with white spots in male and juveniles.

Male has bright red lips and throat during breeding season. Female brown or grey on flanks. Ventral area creamish. **DISTRIBUTION** Widespread. Found in Cambodia, Laos, Thailand, Peninsular Malaysia and Vietnam. Also Bangladesh, Bhutan, India, Nepal, Sri Lanka and Pakistan. **HABITS AND HABITAT** Diurnal. Occurs in deciduous and evergreen forests, but also does well in disturbed areas including plantations and secondary forests.

▪ SKINKS ▪

Keeled Skink ▪ *Eutropis multicarinata* SVL 72mm

DESCRIPTION Medium-sized skink. Olive-green to olive-brown dorsum with narrow dark brown to black vertebral stripe, which runs from nape to forelimbs or slightly more posteriorly. Series of dark spots along dorsolateral margin anteriorly. Lateral surfaces dark with lighter stripe along dorsal edge. Solid or broken stripe from labials to forelimbs. Ventral area bluish or greyish, with small dark spots on chin and throat. **DISTRIBUTION** Previously thought to be widespread, but now considered endemic to the Philippines (Mindanao, Agusan del Sur Province, Dinagat, Samar and Leyte Islands.) **HABITS AND HABITAT** Diurnal. Found among leaf litter on the forest floor, rocks on stream banks, and often under bark and debris in primary, second-growth and midmontane forests. Has also been found in natural bonsai forests. Occurs from sea level to about 400m asl.

Common Sun Skink ▪ *Eutropis multifasciata* SVL 137mm

DESCRIPTION Highly variable, robust-looking skink with short snout. Dorsum generally bronze or brown with usually either yellow or red stripe on flanks. May have white spots or streaks on flanks. Male has bright orange or reddish flanks during breeding season. Five or seven dark lines on dorsal surface, characteristic for this species. Subspecies on Bali (*E. m. balinensis*) has reddish-brown streak on snout and yellow flank-stripes. **DISTRIBUTION** Widespread. Occurs in Brunei Darussalam, Cambodia, Indonesia (*E. m. balinensis* on Bali, *E. m. multifasciata* on remaining islands), Laos, Thailand, Singapore, Malaysia, the Philippines and Vietnam. **HABITS AND HABITAT** Very common diurnal species. Often seen basking in the sun along forest tracks or on tree trunks. Found in forest edges and around human settlements to 1,800m asl.

▪ SKINKS ▪

Rough-scaled Brown Skink ▪ *Eutropis rudis* SVL 120mm

DESCRIPTION Relatively large, robust-looking skink. Dorsum olive-brown with light-edged darker stripe along lateral sides of head and body. Dorsal scales have three strong keels. Male reddish on throat; creamish in female. Throat can have black spots. Ventral area greenish-white. **DISTRIBUTION** Found in Brunei Darussalam, Indonesia (Sumatra, Mentawai Islands, Kalimantan and Sulawesi) and the Philippines (Sulu archipelago). Probably also Malaysia (Sarawak).

HABITS AND HABITAT Diurnal and terrestrial. Mainly found basking in the sun near forest clearings, and similar in habits to other *Eutropis* species. Inhabits lowland primary and secondary forests, but can also adapt to disturbed habitats. Found to 700m asl.

Rough-scaled Sun Skink
▪ *Eutropis rugifera* SVL 65mm

DESCRIPTION Relatively small, robust-looking sun skink with dark or bronze-brown dorsum. Dorsal scales have five (rarely seven) strong keels. Lateral sides lighter in colour, with 5–7 creamish stripes on flanks. Stripes can be broken up into spots. Narrow postocular stripe towards tail. Throat spotted and bright red in male. Ventral side creamish. **DISTRIBUTION** Widespread. Occurs from southern Thailand (Narathiwat, Bannang Sata and Yala Provinces), Malaysia (including Sabah and Sarawak), Brunei Darussalam and Singapore, to Indonesia (Sumatra, Mentawai archipelago, Kalimantan, Java and Bali). Also on Nicobar Islands (India). **HABITS AND HABITAT** Diurnal, terrestrial and relatively uncommon. Often seen basking on trunks. Found in lowlands, in forested areas including primary or secondary forests, swamp forests and forest edges. Can be sighted in low vegetation 1–2m above the forest floor.

▪ SKINKS ▪

Heyer's Isopachys ▪ *Isopachys anguinoides* SVL 95mm

DESCRIPTION Elongated, limbless skink with pale grey-brown dorsum and darker streaks visible on both vertebral and dorsolateral areas. Ventral characterized by brown lines separating scales. **DISTRIBUTION** Endemic to Thailand (Phetchaburi, Surat Thani, Prachuap Khiri Khan and Chumpton Provinces). **HABITS AND HABITAT** Fossorial. Occurs in or under leaf litter debris in lowland forests, and near entrances of caves. Has also been found in open forests and on a golf course near the coast. Reportedly common, but difficult to detect.

Emerald Green Tree Skink ▪ *Lamprolepis smaragdina* SVL 107mm

DESCRIPTION Moderate-sized skink with various morphs such as bright green, green with black blotches on body, and green anteriorly and brown posteriorly. Head pointed; tail almost twice as long as body, and scales smooth. **DISTRIBUTION** Found in Indonesia, the Philippines and Timor-Leste. Nominate form *L. s. smaragdina* occurs in Indonesia and Timor-Leste; subspecies *L. s. moluccarum* in Halmahera Island, Indonesia; *L. s. philippinica* throughout the Philippines. **HABITS AND HABITAT** Conspicuous arboreal skink commonly seen on tree trunks in gardens and other disturbed areas. Often seen in head-down position by day as it actively forages for insects and feeds on flower nectar. Oviparous, laying two eggs per clutch in tree hole.

▪ Skinks ▪

Common Striped Skink
▪ *Lipinia vittigera* SVL 43mm

DESCRIPTION Relatively small, slender skink with long tail. Subspecies *L. v. kronfanum* smaller than *L. v. vittigera*. Brownish-black dorsum with bright yellow vertebral stripe starting at snout-tip. Flanks have dark and pale spots. *L. v. vittigera* has one light stripe; *L. v. kronfanum* five light stripes. **DISTRIBUTION** *L. v. vittigera* found in Cambodia (Mentawai archipelago, Sumatra and Borneo), Laos, Malaysia, Myanmar and Thailand. Subspecies *L. v. kronfanum* found in Vietnam. **HABITS AND HABITAT** Diurnal. Often seen on tree trunks and buttresses, or under tree bark. Inhabits lowland forests and open areas.

Angel's Supple Skink ▪ *Lygosoma angeli* SVL 100mm

DESCRIPTION Slender skink with tail as thick as head. Dorsal colour unpatterned brown; each scale has small black mark. Ventral area creamish with small black spots. **DISTRIBUTION** Found in Cambodia, Laos, Thailand and southern Vietnam (Dong Nai Province). **HABITS AND HABITAT** Diurnal and subfossorial. Often hides under tree bark or in underground burrows. Inhabits mainly lowland forests in riparian habitats.

▪ Skinks ▪

Annamese Supple Skink ▪ *Lygosoma corpulentum* SVL 170mm

DESCRIPTION Large, very robust skink, with yellowish- or chocolate-brownish dorsum. Dense mottled pattern on dorsum and lateral sides. Supralabials and infralabials yellowish with black edges. Throat and neck-sides yellow. **DISTRIBUTION** Found in southeastern Thailand (Chachaengsao and Chantaburi Provinces), Laos (Champasak Province) and Vietnam (Da Lat, Lam Dong Province). **HABITS AND HABITAT** Habits little known. Probably similar to other supple skinks (diurnal and semi-fossorial). Occurs in wet montane forests.

Korat Supple Skink ▪ *Lygosoma koratense* SVL 106mm

DESCRIPTION Robust, elongated-looking skink. Dorsum reddish-brown with black tips on scales. Flanks greenish and lateral scales have similar black-tipped scales. Scales on head have darker edges with spot in middle, and yellow with black-spotted labials. Ventral area creamish. **DISTRIBUTION** Found in central and eastern Thailand (Nakhon Ratchasima, Korat and Saraburi Provinces). **HABITS AND HABITAT** Very little known. Appears to be subfossorial and inhabits lowland forests near bases of limestone mountains, occurring under fallen trees and tree logs.

▪ SKINKS ▪

Short-limbed Supple Skink ▪ Lygosoma quadrupes SVL 96mm

DESCRIPTION Very elongated skink with reduced limbs and tail as thick as body. Dorsum yellowish-brown, with thin dark lines towards tail. Forehead and supralabials darker than dorsum. Ventral area and subcaudals pale pink. **DISTRIBUTION** Restricted to Java (Indonesia). **HABITS AND HABITAT** Subfossorial species often found in or near leaf litter. Folds back reduced limbs along body during rapid movement. Found in lowland forests, open areas (forest edges and agricultural land) to 700m asl.

Philippine Giant Skink ▪ Otosaurus cumingii SVL >115mm

DESCRIPTION Large, robust skink that is the only forest skink in the region with large supranasal scales and larger body than other related skinks. Dorsum and head brownish with black markings resulting in incomplete transverse bands. Black band from eye towards tail, frequently interrupted by yellowish spots, sometimes so close together that they form vertical bands. Ventral area greyish or creamish. Legs have marbled pattern. **DISTRIBUTION** Found in the Philippines (Mindanao, Bohol, Dinagat, Luzon, Mindoro, Calotcot, Sibuyan and Sicogon). **HABITS AND HABITAT** Often seen foraging under leaves and logs, or in open spaces by day. Occurs at 200–500m asl.

◼ SKINKS ◼

Doria's Ground Skink ◼ *Scincella doriae* SVL 59mm

DESCRIPTION Robust-looking skink with brownish dorsum. Small, dark brown spots and dark lateral stripe from nostril to flanks. Lateral stripe broken up by paler looking spots towards flanks. Ventral area yellowish or white. DISTRIBUTION Found in northeastern Myanmar (Kakhien Hills, Bahmo, Kachin State), northeastern Thailand (Doi Inthanon and Doi Suthep, Chiang Mai Province and Loei Province) and Vietnam. Also southern China. HABITS AND HABITAT Diurnal and terrestrial. Often found in or near leaf litter. Inhabits primary forests from mid-hills towards submontane limits.

Black Ground Skink ◼ *Scincella melanosticta* SVL 57.4mm

DESCRIPTION Robust-looking skink with relatively small head and long limbs. Dorsum olive, bronze or golden-brown, with large brown or black spots along vertebral region. Thin stripe from nostril towards flanks, where it widens and is broken up by pale spots. Lower flanks pale white with small black spots. Supralabials and infralabials creamish with black sutures. Male has bright orange head, neck and throat during breeding season. DISTRIBUTION Widespread. Found in Cambodia (Cardamom Mountains), Thailand (Upper Mekong, Dong Phaya Fai Range, Huey Sapon, Nakon Sri Thamarat and Chanthabun), eastern Myanmar (Mt Mulayit and Tenasserim, Tanithayi Division) and Vietnam (Langbian Plateau, Lam Dong Province). HABITS AND HABITAT Diurnal and terrestrial. Often seen near rocks and fallen logs. Inhabits evergreen forests at 100–1,220m asl.

◼ SKINKS ◼

Reeves's Ground Skink ◼ *Scincella reevesii* SVL 57.4mm

DESCRIPTION Typical *Scincella* skink with slender, elongated body. Dorsum bronze-brown, with black spots concentrated in vertebral region. Dark brown stripe that is narrow on head, but widens towards flanks, where it is broken up by cream spots. Ventral area cream to yellowish, with no pattern.

DISTRIBUTION Found in north and south-east Myanmar, Thailand and Vietnam. Possibly also Laos and Cambodia. HABITS AND HABITAT Diurnal and terrestrial. Ovoviparous, laying clutches of 2–5 eggs. Displays strong association with rocky habitats, where it can be found in forested hills to 1,500m asl.

Blue-throated Forest Skink
◼ *Sphenomorphus cyanolaemus* SVL 60.2mm

DESCRIPTION Slender-looking skink with golden head, changing to bronze towards tail. Tail black with blue mottled pattern. Male has deep blue head, throat and pectoral region; lighter blue in female. Female and juveniles show more black and less intense blue colours. Dark postorbital stripe towards just above tympanum.

DISTRIBUTION Found in Indonesia (Sumatra and Borneo) and Peninsular Malaysia (Perak, Pahang and Selangor). Probably occurs throughout Peninsular Malaysia. HABITS AND HABITAT Diurnal, and terrestrial to semi-arboreal. Often seen on logs, small vegetation and low tree trunks. Frequently sighted foraging on the ground in direct sunlight. Often found sleeping on leaves or saplings overhanging hill streams. Occurs in lowland dipterocarp forests to 300m asl.

∎ SKINKS ∎

Banded Sphenomorphus ∎ *Sphenomorphus fasciatus* SVL 80.7mm

DESCRIPTION Morphologically distinct species most closely related to Solomon Island taxa. Long and slender, with tail longer than body. Dark brown base colour all over body; body covered in spots that are so closely placed together on dorsum that they form transverse bands. Spots yellowish or brownish on head and dorsal area, but whitish on flanks and tail. DISTRIBUTION Endemic to the Philippines (Mindanao, Bohol, Camiguin Sur, Dinagat, Samar and Leyte Islands). HABITS AND HABITAT Diurnal. Found under leaves and rotten logs, and between tree buttresses in dipterocarp and submontane forests. Common in low-elevation, disturbed second growth and coastal forests. Also found in overhangs of cliffs. Occurs from sea level to 1,200m asl.

Indian Forest Skink ∎ *Sphenomorphus indicus* SVL 97mm

DESCRIPTION Slender-looking skink with head easily distinguished from neck. Dorsal area of head, body and tail golden-brown, with or without pattern or brown spots. Dark brown lateral stripe from nostril to tail-base, fading towards lower flanks. Flanks can have orange shade and brown mottled pattern. Lateral stripe almost black and bordered with white at bottom in juveniles. DISTRIBUTION Widespread. Found in Cambodia, Laos, Myanmar, Peninsular Malaysia, Thailand and Vietnam. HABITS AND HABITAT Diurnal and terrestrial. Often found among leaf litter or fallen trees in rocky habitat. Frequently associated with uphill ecosystems throughout its range, although in Peninsular Malaysia it is restricted to hill dipterocarp and montane forests above 1,100m asl. In other areas found in lowland evergreen forests to submontane forests to 1,250m asl.

▪ SKINKS ▪

Gunung Kinabalu Skink ▪ *Sphenomorphus kinabaluensis* SVL 58mm

DESCRIPTION Dorsum olive to dark brown, with black dorsolateral stripe postocular over axilla and along dorsolateral area. Dorsum has longitudinal rows of brown to yellow spots with occasional brown speckles. Ventral area creamish. Gular area can have black spots or speckles, with labials having narrow brown or black sutures. Tail similar in colour to dorsum, but underside can be red/rose fading to blue. Similar to the Blue-throated Forest Skink (p. 119), but trunk more elongated in this species, and Blue-throated lacks black spots on gular area. **DISTRIBUTION** Endemic to Malaysia (Gunung Kinabalu and Crocker Range, Sabah). **HABITS AND HABITAT** Habits very little known, but probably diurnal. Oviparous, with clutches of 1–2 eggs. Inhabits submontane and montane forest at 1,600–2,200m asl.

Line-spotted Forest Skink
▪ *Sphenomorphus lineopunctulatus* SVL 84mm

DESCRIPTION Slender-looking skink with short snout. Dorsum has olive-brownish base. Forehead darker than rest of body, and has tiny black spots. Axilla has black spots. Flanks have dark stripe about 2–3 scales wide. Grey below dark stripe. Neck, throat and lower jaw greyish. Ventral area creamish. **DISTRIBUTION** Found in Cambodia, Laos and Thailand. **HABITS AND HABITAT** Habits very little known. Originally found by day on sparsely overgrown rocky plain, close to disturbed semi-evergreen forests or under stones in secondary forests.

■ SKINKS ■

Maculated Forest Skink ■ *Sphenomorphus maculatus* SVL 67mm

DESCRIPTION Slender-looking skink with bronze-brown dorsum, which can be unpatterned or have paravertebral spots from nape to tail-base. Dark lateral stripe bordered on ventral side by thin white stripe, and on dorsal side by light brown spots. Lateral stripe from nostril, through eye, to tail-base. Head, neck and flanks have dark irregular markings. Legs dark with mottled pattern. **DISTRIBUTION** Occurs in Cambodia, Laos, Myanmar, Thailand and Vietnam. Probably also Peninsular Malaysia, but reports unconfirmed to date. **HABITS AND HABITAT** Diurnal and terrestrial. Most commonly seen by day foraging along streams or in leaf litter. Found in lowlands, primary evergreen forests and open habitats from mangrove swamps to 800m asl.

Blotched Forest Skink
■ *Sphenomorphus praesignis* SVL 121mm

DESCRIPTION Smooth-scaled, robust-looking skink with short snout. Ground colour of body light brown to brown-orange. Labials have black and white bands. Three to five black spots cover anterior region of flanks, slowly breaking up and creating speckled pattern. Tail dark, with lighter bands or white spots. Ventral area unpatterned. Juveniles have much more intense black mottled pattern. **DISTRIBUTION** Found in southern Thailand (south of Isthmus of Kra) and Peninsular Malaysia (Banjaran Timur, Banjaran Bibntang and Banjaran Titiwangsa). **HABITS AND HABITAT** Secretive diurnal species rarely seen moving around, but can often be seen in crevices, hollow logs and rock cracks. Usually sighted sticking its head out of a crack. Inhabits hill dipterocarp and lower montane forests at 820–1,280m asl.

▪ SKINKS ▪

Sabah Forest Skink ▪ *Sphenomorphus sabanus* SVL 58mm

DESCRIPTION Robust, colourful skink with dorsal colour ranging from pinkish-brown to olive-yellow, with darker and light mottled pattern. Forehead orange-brown with dark grey markings. Sides of neck and flanks of male have orange shade on flanks. Labials barred with black. Scales around eye yellow, simulating yellow eye-ring. DISTRIBUTION Endemic to Borneo (Sabah, Malaysia). HABITS AND HABITAT Diurnal and mostly arboreal. Ovoviparous, laying clutches of 2–3 eggs. Often sighted climbing tree trunks and buttresses, but also found among leaf litter. Locally common and found in damp lowland rainforests to submontane forests to 850m.

Selangor Forest Skink
▪ *Sphenomorphus scotophilus* SVL 50mm

DESCRIPTION Slender skink with obtusely pointed snout and large eyes. Dorsum dark brown with dark and yellowish spots. Dorsolateral area has series of rounded cream spots. Labials patterned with white and black spots. Ventral area creamish with no pattern. DISTRIBUTION Found in southern Thailand (Khao Chong, Trang Province) and Peninsular Malaysia (Pulau Aur, Pulau Pemanggil, Pulau Tioman, Pulau Tulai, Seberang Peri, Pulau Penang and Kepong). HABITS AND HABITAT Diurnal and arboreal. As a scansorial species, associated with tree trunks and rocks, mainly feeding on small arthropods. Inhabits lowland forests and limestone cave ecosystems.

▪ SKINKS ▪

Thai Forest Skink ▪ *Sphenomorphus tersus* SVL 96mm

DESCRIPTION Relatively large skink with reddish-brown base colour. Dark, faded and diffuse markings on top of head. Labials have banded pattern; faded dorsolateral markings on body from nape towards groin. Tail thick, darker in colour than body and has little or no pattern. Juveniles have more bold and more contrasting pattern than adults. **DISTRIBUTION** Occurs in Thailand (Isthmus of Kra) and Peninsular Malaysia. **HABITS AND HABITAT** Shows a great affinity with water and often found among boulders or leaf litter near streams. Inhabits lowlands in primary and secondary dipterocarp forests below 200m asl.

Variable Forest Skink ▪ *Sphenomorphus variegatus* SVL 62.9mm

DESCRIPTION Robust-looking skink with brown to reddish base colour; tail and hindlegs darker and gradually turn lighter towards front. Grey or silver spots on sides. Clear black spot on flanks just behind head, surrounded by several lighter coloured, smaller spots. Jaws and head have black markings. **DISTRIBUTION** Found in the Philippines (Mindanao, Samar, Leyte and Bohol). Also Malaysia (Borneo) and Indonesia (Sulawesi), although this is uncertain. **HABITS AND HABITAT** Typical *Sphenomorphus* in habits. Primarily found in areas with relatively dense and thick canopy cover, or next to small creeks and shaded parts of stream beds. Also occurs where mature second- and first-growth forests are present. Low-elevation species, found at 300–400m asl.

▪ SKINKS ▪

Bowring's Supple Skink ▪ *Subdoluseps bowringii* SVL 58mm

DESCRIPTION Slender skink with head nearly indistinct from neck, and thick, rounded tail. Bronze-brown dorsum with black band on flanks, with white and black spots that may form longitudinal lines. Juveniles have bright red tail compared to grey tail in adults, and unpatterned yellow ventral area. **DISTRIBUTION** Widespread.

Found in Cambodia, Indonesia (Kalimantan, Java, Bali, Sumatra, Pulau Berhala and Pulau Weh), Thailand, Laos, Peninsular Malaysia, Singapore, and the Philippines (Sulu archipelago, introduced to Mindanao). Probably represents a species complex. **HABITS AND HABITAT** Diurnal and fossorial. Inhabits open areas and clearings in plains, evergreen forests and plantations in mid-hills. Also found in towns and cities.

Herbert's Supple Skink ▪ *Subdoluseps herberti* SVL 67mm

DESCRIPTION Slender skink with pointed snout. Scales have five keels. Dorsum bronze-brown. Dark postocular stripe towards axilla, continuing on flanks. Sides of neck

and tail have pale scales, looking like pale spots. Ventral area light brown. **DISTRIBUTION** Found in southern Thailand (Ranon, Phang-nga, Phuket and Surat Thani Provinces, and Khao Wang Hip, Nakon Si Thammarat Province) and northern Peninsular Malaysia (Sungai Menora and Kangsar, Perak State). **HABITS AND HABITAT** Little known, but probably similar to other supple skinks. Occurs near mountain foothills.

▪ SKINKS ▪

Sama Jaya Supple Skink ▪ *Subdoluseps samajaya* SVL 70.1mm

DESCRIPTION Slender, elongated skink with brown base colour on dorsal surface, body, limbs and tail. Lips cream with brown markings. Gular and ventral surfaces creamish. Dark brown lateral stripe from nostril, through eye and fading towards brown dorsum after front legs. Light spots visible laterally. DISTRIBUTION Known from several locations in western Sarawak (Malaysia), including the Sama Jaya Forest Reserve, Kuching. HABITS AND HABITAT Assumed to be semi-fossorial in habits; has been found among leaf litter in dense, closed-canopy lowland rainforests.

Red-eyed Crocodile Skink ▪ *Tribolonotus gracilis* SVL 97mm

DESCRIPTION Robust skink with large, triangular head bearing bony crest, and characteristic red-pigmented semi-circle around eye. Four rows of backwards-pointed spikes from nape, gradually reducing in size after hindlegs. Base colour dark brown. Male has section of enlarged belly scales just in front of hindlegs, giving 'six-pack' effect. Toe-pads visible on third and fourth toes in male. DISTRIBUTION Found in Papua and West Papua Provinces (Indonesia). HABITS AND HABITAT Secretive skink; one of the few skinks that vocalize loudly when disturbed. Often found in or near streams, under rotten logs or in leaf litter; also occurs in coconut-husk piles near coconut plantations.

▪ Skinks ▪

Beccari's Water Skink ▪ *Tropidophorus beccarii* SVL 98mm

DESCRIPTION Robust-looking, smooth-scaled skink when fully grown; juveniles look slightly more slender. Dorsum dark or reddish-brown, with dark blotches and cross-bars visible. Sides of head and flanks have light spots and dark stripe with lighter coloured, vertical stripe-like markings visible. **DISTRIBUTION** Endemic to Borneo (Brunei Darussalam, Indonesia and Malaysia). **HABITS AND HABITAT** Inhabits rocky streams in dipterocarp forests to 1,000m asl. Ovoviviparous, giving birth to four live young.

Brooke's Water Skink ▪ *Tropidophorus brookei* SVL 101mm

DESCRIPTION Large, robust-looking skink with rounded body, although juveniles are more slender than adults. Dorsum dark brown with darker blotches and spots; the markings form indistinct transverse bands. Both sides of neck have black marking; black and white markings on flanks. **DISTRIBUTION** Found in Borneo (Brunei Darussalam, Indonesia and Malaysia). **HABITS AND HABITAT** Typical *Tropidophorus* species in habits, and thus diurnal and semi-aquatic. Inhabits forested hills at 600–900m asl, where probably restricted to habitats near streams.

◾ Skinks ◾

Indo-Chinese Forest Skink ◾ *Tropidophorus cocincinensis* SVL 80mm

DESCRIPTION Robust, rounded-looking skink. Scales keeled in juveniles, smooth in adults. Caudal scales strongly keeled. Dorsum light brown with indistinct darker markings. Flanks brownish to black, with small white speckles and larger orange or red spots/markings. Gular area has grey mottling pattern; ventral area creamish. Labials light coloured and barred. **DISTRIBUTION** Found in Thailand (Khao Phanom Dongrak Range), Cambodia, Laos (Xe Kong Province) and Vietnam (Kon Tum, Quang Binh, Thua-Thien-Hue, Da Nang and Thua Thien Hue Provinces). **HABITS AND HABITAT** Diurnal and semi-aquatic. Ovoviviparous, with litters comprising 7–9 hatchlings. Primarily occurs near rocky streams in lowland forests. Found at 109–160m asl.

Philippine Spiny Water Skink ◾ *Tropidophorus grayi* SVL 102mm

DESCRIPTION Primarily characterized by spinose dorsal body scales. Dorsum base colour greyish-black, but may vary from brown to black. Ventral surface white and can have brown blotches. May have several lighter brown lateral stripes from back of neck to tail-tip. **DISTRIBUTION** Endemic to the Philippines (Luzon, Polillo, Masbate, Panay, Negros, Cebu, Catanduanes, Leyte and Samar Islands). **HABITS AND HABITAT** Diurnal but secretive in habits, hiding in the water when threatened. Semi-aquatic, foraging in the leaf litter along streams or in shallow water. Mainly found in crevices in clay or rocky banks along streams, but also under moist humus and logs on the forest floor or near forest streams to 800m asl. Appears to be abundant, and also found in secondary forests and (human) disturbed areas.

▪ Skinks ▪

Laotian Water Skink ▪ *Tropidophorus laotus* SVL 75mm

DESCRIPTION Robust, rounded-looking skink with smooth scales. Front nasal and often also anterior loreal divided. Dorsum dark brown, with occasionally wedge-shaped markings that are light in colour with black edges. Flanks have small white spots; ventral area creamish. Subcaudal scales have black spots. **DISTRIBUTION** Found in Thailand (Loei and Nong Khai Provinces) and Laos (Muang Liep, Sainyabuli Province). **HABITS AND HABITAT** Crepuscular or nocturnal. Semi-aquatic and strongly associated with hill streams. Occurs in submontane forests. Ecology little known.

Small-scaled Water Skink ▪ *Tropidophorus microlepis* SVL 83mm

DESCRIPTION Relatively large, robust-looking skink characterized by strongly keeled, spinous dorsal and lateral scales. Scales on forehead have a rough look to them. Series of small scales visible between loreals and supralabials. Dorsum light brown with six dark brown blotches. Indistinct blotches on flanks, and both limbs and digits banded. Anterior lateral stripe is present. Ventral area creamish and unpatterned. **DISTRIBUTION** Found in southeastern Thailand (Chataburi and Sa Kaeo Provinces), Laos (Champasak Province), eastern Cambodia and central Vietnam (Lam Dong Province). **HABITS AND HABITAT** Diurnal and semi-aquatic, like other *Tropidophorus* species. Ovoviviparous, giving birth to 7–9 hatchlings. Occurs in primary evergreen forests at 180–440m asl.

▪ SKINKS ▪

Small-legged Water Skink ▪ *Tropidophorus micropus* SVL 40mm

DESCRIPTION Relatively slender-looking skink with striated scales on forehead. Dorsum dark brown with black blotches. Flanks and neck-sides creamish-white with irregular black markings. Tail similar in colour to dorsum. Labials white with no pattern. **DISTRIBUTION** Endemic to Borneo (Sarawak and Sabah, Malaysia). **HABITS AND HABITAT** Diurnal and semi-aquatic. Ovoviviparous, giving birth to three hatchlings. Found near rocky streams in lowland forests, where it hides in fallen tree trunks and rock crevices, and even under waterfalls. Excretes musk when disturbed.

Misamis Water Skink ▪ *Tropidophorus misaminius* SVL 112mm

DESCRIPTION Large, slender skink with strongly keeled dorsal scales. Dorsum brown with traces of blackish postocular streak. Cluster of white spots on temples, with a few additional scattered white spots on neck. Ventral area whitish; throat, neck, palms, soles and posterior two-thirds of tail blackish. **DISTRIBUTION** Endemic to the Philippines (Mindanao, Camiguin and Basilan Islands). **HABITS AND HABITAT** Semi-aquatic. Usually occurs near or in rocky beds and on stream banks. Occasionally found under rotten logs or on banks of cave streams.

▪ Skinks ▪

Partello's Water Skink ▪ *Tropidophorus partelloi* SVL 126.5mm

DESCRIPTION Relatively large skink with moderately tapered snout. Dorsum dark to dusky brown, with head usually slightly lighter in colour; 7–9 irregular, narrow and light transverse bands of 1–2 scale rows on dorsum. Ventral area creamish to greyish. Both dorsal and lateral scales have low keels, with scales on tail raised to sharp spines posteriorly. DISTRIBUTION Endemic to the Philippines (Mindanao). HABITS AND HABITAT Semi-aquatic. Ovoviviparous. Found in damp soil under logs or rocks in forested habitats 450–1,200m asl.

Robinson's Water Skink ▪ *Tropidophorus robinsoni* SVL 75mm

DESCRIPTION Robust-looking skink with dark brown to almost blackish dorsum; light-coloured, black-edged bands on both dorsum and tail. Ventral surfaces white to creamish. Both gular and ventral sides of tail can have black spots. Labials have white to creamish spots. DISTRIBUTION Occurs in Myanmar (Tanintharyi Division) and Thailand (Chumpon, Phang-nga and Surat Thani Provinces). HABITS AND HABITAT Diurnal and semi-aquatic. Strong association with streams, like other *Tropidophorus* species. Primarily found in lowland and dry-evergreen forests.

Baleh Water Skink ■ *Tropidophorus sebi* SVL 85.9mm

DESCRIPTION Robust-looking skink with brownish dorsum; darker transverse bars visible on dorsum and tail. Dark postocular stripe that narrows along flanks of torso. Flanks greyish with darker markings and yellow to orange flecks. Tail-sides dark, with greyish spots or flecks. Toes brown with darker bands. Head brown with no markings; temporal region and labials have grey-black banding or markings. **DISTRIBUTION** Occurs in Putai, Upper Baleh, Sarawak, Malaysia. **HABITS AND HABITAT** Found along banks of small streams, in or near narrow crevices among rocks. Ecology little known, but probably similar to that of other *Tropidophorus* species.

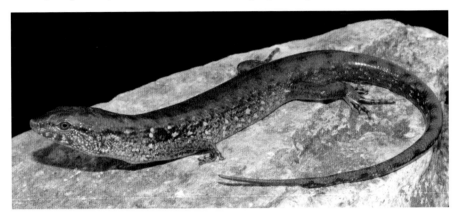

Chinese Water Skink ■ *Tropidophorus sinicus* SVL 65mm

DESCRIPTION Relatively small, robust-looking skink with narrow head. Scales on forehead strongly striated. Dorsum dark brown; 10 lighter brown bars on dorsum and tail, edged by darker scales. Dorsal scales strongly keeled. Labials dark with irregular light markings. Creamish spots on flanks. **DISTRIBUTION** Found in Vietnam (Man Son Mountains), possibly also in Thailand (Nan Province) and Laos. **HABITS AND HABITAT** Probably nocturnal and semi-aquatic, showing strong association with stream habitats. Ovoviviparous, giving birth to up to six neonates. Primarily found in montane forests.

▪ Skinks ▪

Cursed-stone Diminutive Leaf-litter Skink
▪ *Tytthoscincus batupanggah* SVL 33.2mm

DESCRIPTION Relatively small skink with slender body and smooth scales. Dorsal and lateral sides have brown base colour with cream-coloured spots across body. Cream dorsolateral strip from orbit to midway on body, underlined by brown lateral stripes, both fading towards hindlimbs. Orange colouration visible near front limbs, on both lateral and dorsal sides. DISTRIBUTION Currently only known from Batu Panggah on north side of Gunung Penrissen, Sarawak, Malaysia. May occur throughout high-elevation forests on Gunung Penrissen. HABITS AND HABITAT Known to occur in primary, highland mixed-dipterocarp forests, where primarily found in and around leaf litter. Only known individuals found at 1,050m asl.

Hallier's Forest Skink ▪ *Tytthoscincus hallieri* SVL 57mm

DESCRIPTION Slender-looking, smooth-scaled skink. Characteristic temporal scale between last supraocular and parietal. Dorsum brown with reddish-brown postocular stripe behind axilla. Stripe breaks up into spots or stripe-like markings behind trunk. Flanks light brown or olive, with small, yellow or greenish spots. Gular area of male blue with black markings; gular area in female pink or yellow without black markings. DISTRIBUTION Endemic to Borneo (Kalimantan, Indonesia and Sabah, Malaysia). HABITS AND HABITAT Ecology little known. Probably diurnal in habits, similar to other *Tytthoscincus* species. Known to inhabit lowlands and mid-hills in dipterocarp forests.

▪ SKINKS ▪

Singapore Swamp Skink ▪ *Tytthoscincus temasekensis* SVL 36.3mm

DESCRIPTION Relatively small skink with slender body and smooth scales. Base colour of head, body, limbs and tail dark brown, fading to beige on ventral surfaces. Head has light spots; labials barred and with light-coloured spot. Dorsal pattern either more spotted or lineated, and faded in some individuals. Differentiated from other montane *Tytthoscincus* by having stripes or longitudinal spots, and from swamp *Tytthoscincus* by having two versus one loreal scales. **DISTRIBUTION** Occurs in lowland areas in Singapore and in vicinity of Tanjung Malim, Perak, Peninsular Malaysia. **HABITS AND HABITAT** Found in lowland dipterocarp forests, along stream banks in peat and freshwater swamp forests, occurring at to 37m asl. Individuals have been caught accidentally during fishing and seen at edge of streams, suggesting a semi-aquatic habit.

■ CROCODILE LIZARDS ■

SHINISAURIDAE (CROCODILE LIZARDS)
The crocodile lizards consist of only one species and two subspecies, of which one subspecies is restricted to Southeast Asia. These semi-aquatic lizards are diurnal and secretive, and are threatened by habitat loss and poaching.

Vietnamese Crocodile Lizard
■ *Shinisaurus crocodilurus vietnamensis* SVL 300mm

DESCRIPTION Robust lizard with very short head. Dorsum grey or reddish-brown. Red, orange and yellow blotches on head, throat and lateral sides (male). Male has enlarged scales on neck and larger cranial crest compared to female. Dark radiating pattern around eye. Tail banded with two rows of enlarged scales. **DISTRIBUTION** Vietnamese subspecies found in northeastern Vietnam (Yen Tu Nature Reserve, Quang Ninh Province). **HABITS AND HABITAT** Diurnal and arboreal. Ovoviviparous, giving birth to 2–15 young. Often found sleeping on branches above shallow ponds and streams. Can be seen by day foraging near water. When threatened, will try to escape via the water; can remain submerged for up to 30 minutes. Occurs mainly in evergreen forests at 200–1,500m asl. Strongly associated with water, and mainly found in dense vegetation surrounding water.

■ Monitor Lizards ■

Varanidae (Monitor Lizards)

The Varanidae are distributed throughout Africa, the Middle East, Asia, Australia and Melanesia. They range from 116mm SVL to an astonishing 1.5m SVL in the Komodo Dragon, the largest lizard species in the world. At the time of writing 40 monitor lizard species occur in, or are completely restricted to, Southeast Asia. Monitor lizards inhabit a wide range of habitats both terrestrial and arboreal, and can also be semi-aquatic in nature.

Bengal Monitor ■ *Varanus bengalensis* SVL 610mm

DESCRIPTION Large, robust monitor with variable colours and patterns, depending on localities. Dorsum brown to black with many light or dark scales. Ventral body lighter brown with darker speckles. Nostrils slit shaped in adults and oval in juveniles. Nostril nearer eye than snout-tip. Tail laterally compressed, with keel. Juveniles dark grey with 11 narrow transverse light yellow bands (1–3 scales wide) on dorsal body. **DISTRIBUTION** Found in Myanmar. Once thought to be widely distributed from Southeast to south Asia. Restricted to Myanmar in Southeast Asia after the Clouded Monitor (p. 141) was elevated to species status. **HABITS AND HABITAT** Diurnal and primarily terrestrial. Feeds on small animals such as land snails, insects, arachnids, centipedes and frogs.

▪ Monitor Lizards ▪

Northern Sierra Madre Forest Monitor
▪ *Varanus bitatawa* SVL 770mm

DESCRIPTION Large-bodied monitor with black base colour, and yellowish-green dots and spots. At least four rows of ocelli on body; head robust; nostril opening a slanted slit; well-developed, muscular limbs with curved claws. Tail has alternating broad, black and yellowish-green bands. One of three members of subgenus *Philippinosaurus*, a clade of the only known frugivorous monitor lizards in the world. **DISTRIBUTION** Found in the Philippines (Luzon). **HABITS AND HABITAT** Highly secretive, arboreal and forest obligate. Diet includes fruits of *Pandanus*, *Canarium*, and *Ficus*, and invertebrates such as land snails and hermit crabs. Most active during late morning to early afternoon. A telemetry study showed males with greater activity areas (23km^2) than females (5.2km^2). Oviparous; no other information currently known about its reproductive biology.

■ Monitor Lizards ■

Mindanao Water Monitor ■ *Varanus cumingi* SVL 600mm

DESCRIPTION Large monitor with muscular limbs. Overall black and yellow; black temporal stripe. Nostrils oval, near snout-tip. Amount of yellow on head and body variable; transverse rows and spots may fade with age. Mid-body scales 120–150. DISTRIBUTION Found in the Philippines (Basilan, Mindanao and Siargao). HABITS AND HABITAT Terrestrial and diurnal. Young individuals mainly feed on insects; adults feed on fish, crabs, molluscs and carrion. Predators include birds of prey and the King Cobra *Ophiophagus hannah*. Often seen in disturbed areas such as farmland and forest edges.

Dumeril's Monitor ■ *Varanus dumerilii* SVL 500mm

DESCRIPTION Moderate-sized brown monitor with 4–5 lighter bands on dorsal body. Nostrils oval or slit-shaped, closer to eye than snout-tip. Scales on neck round, flat and enlarged. Dark band behind eye to upper back; 3–5 vertical stripes run across mouth. Colour of juveniles more contrasting than in adults; head and neck bright orange; dorsal body and tail have 10–12 pale yellow to yellowish-orange, transverse stripes. Colour fades with age. DISTRIBUTION Found in Brunei Darussalam, Indonesia, Malaysia (Peninsular and Borneo), Myanmar, Singapore and Thailand. HABITS AND HABITAT Diurnal and terrestrial. Feeds on crabs, insects, arachnids, fish and frogs.

■ Monitor Lizards ■

Komodo Dragon ■ *Varanus komodoensis* SVL 1,570mm

DESCRIPTION Largest and most iconic of all monitors, potentially weighing up to 100kg. Head, body and limbs robust. Scales around eyes yellowish to off white. Nostrils oblong and near snout-tip. Adults overall greyish-brown; neck and back of subadults yellowish to rusty-red; juveniles more colourful, with black-grey base colour, 'V'-shaped pattern on neck, yellow to orange/red spots on back and limbs, and banded tail. Ventral area of juveniles whitish to pale yellow with distinctive dark stripes. **DISTRIBUTION** Found in Indonesia on the islands of Flores, Komodo, Rinca, Padar, Nusa Kode and Gili Motang. **HABITS AND HABITAT** Famously known to predate on large animals such as water buffalos and deer, but also feeds on carrion and smaller Komodo Dragons. Occurs in a variety of habitats, including monsoon forests, savannah and steppes, mangrove forests and littoral zones. Highest abundances in savannah and monsoon habitats. Juveniles arboreal for the first year to escape predation by older Komodo Dragons, but feed both on the ground and in trees, primarily on insects. Late maturing (5–7 years; 70cm SVL). Female lays 18.5 eggs per clutch on average. Mainly occurs at low elevations, but also above 800m asl.

Monitor Lizards

Blue Tree Monitor — *Varanus macraei* SVL 360mm

DESCRIPTION Member of the *V. prasinus* complex. Dorsum black with scattered blue scales, spots and ocelii, sometimes arranged in up to nine blue bands. Head mostly blue with labials a whitish colour. Throat bright turquoise with clear 'V' shape on neck. Limbs speckled, forming ocelli on upper legs. Scales on tail uniform in size, with 22–23 symmetrical blue bands on black ground colour. Like other members of *V. prasinus* complex, has a strong prehensile tail almost twice as long as the SVL. **DISTRIBUTION** Currently thought to be endemic to Batanta Island, Indonesia, with one of the smallest distributions of all monitor lizards. Reportedly also occurred on an unknown offshore island off Batanta, but anectodal evidence suggest that the species was lost from this island due to over-collection for the pet trade. Reports also state that it is or was present on Pulau Ayem, where it might occur in very low densities. **HABITS AND HABITAT** Diurnal and arboreal, and found in forested habitats. Reportedly rare, or cryptic, on Batanta Island. Feeds on invertebrates such as insects and spiders. Female lays 3–5 eggs per clutch. Artificially incubated eggs kept at 29–31 °C hatched after 150–183 days. Hatchling 90–100mm SVL; weight 10–15g. **NOTE** This species is threatened by over-collection for the international pet trade due to its high commercial value, and is under increasing threat from deforestation and urban development.

■ Monitor Lizards ■

Philippine Marbled Water Monitor
■ *Varanus marmoratus* SVL 450mm

DESCRIPTION Member of *V. salvator* complex. Black base colour and yellow spots on body and limbs. Snout has alternating light cream and black cross-bands; nostril round near snout-tip; light cream temporal streak between eye and ear opening (tympanum); scales on neck enlarged. Juveniles have 4–6 transverse rows of large ocelli on body, limbs and tail that may vanish with age. **DISTRIBUTION** Endemic to the Philippines (Batan, Lubang, Luzon and Calayan). **HABITS AND HABITAT** Generalist, feeding

on live vertebrates and invertebrates and their carcasses. Documented to feed on the invasive and poisonous Marine Toad *Rhinella marina* with no apparent ill effects. Occasionally considered a pest due to habit of preying on domesticated animals such as chickens and ducks. Oviparous, laying 8–14 eggs per clutch. Typically found in various lowland habitats such as forests, mangroves, coasts, fishponds, farms and coconut groves.

Quince Monitor
■ *Varanus melinus* SVL 500mm

DESCRIPTION Moderate-sized, slender monitor with light yellow head. Nostrils round, near snout-tip. Forked tongue pink or flesh coloured. Body and legs light yellow with dark reticulation. Ventral area light yellow, with or without faint reticulated pattern. Tail has faint dark bands. **DISTRIBUTION** Found in Indonesia (Sula Islands). **HABITS AND HABITAT** Diurnal. Adept both on the ground and in climbing trees. Limited observation in the wild.

■ Monitor Lizards ■

Clouded Monitor ■ *Varanus nebulosus* SVL 610mm

DESCRIPTION Large monitor with robust body, and strong legs and tail. Dorsum ranges from light to dark brown, with many small, light yellow scales and spots. Juveniles darker grey than adults, with yellow spots on body and legs; tail has dark bands (distance between bands wider posteriorly). DISTRIBUTION Found in Cambodia, Indonesia, Malaysia, Myanmar, Singapore, Thailand and Vietnam. HABITS AND HABITAT Diurnal. Mainly feeds on the ground, but known to be agile tree climber. Primarily feeds on invertebrates such as insects, arachnids, ants and snails, and rarely on frogs, fish, lizards, snakes and small mammals.

Western Visayas Water Monitor ■ *Varanus nuchalis* SVL 530mm

DESCRIPTION Moderate-sized monitor with two distinct morphs; one is overall black with or without bright yellow scales on limbs and tail, the other black with varying amount of white on head, and white or yellow oval spots. Both morphs have thin mid-dorsal line and black temporal stripe. Mid-body scales 136–169. DISTRIBUTION Found in the Philippines (Cebu, Masbate, Negros, Panay, Tablas and Ticao). HABITS AND HABITAT Active by day. Opportunistically feeds on invertebrates (insects, crustaceans), vertebrates (birds, rodents, frogs), and carrion. Has been documented feeding on the invasive and poisonous Marine Toad without ill effects. Most common in lowland forests. Often seen in mangroves, fishponds, rice fields and other disturbed areas.

■ Monitor Lizards ■

Southern Sierra Madre Forest Monitor
■ *Varanus olivaceus* SVL 730mm

DESCRIPTION First known member of subgenus *Philippinosaurus*, scientifically described in 1857. Large-bodied, forest-obligate lizard; greenish-grey with dark transverse bands on neck and dorsal body. Nostrils slit shaped. Scales around nostrils, eyes and labials may be bright yellow. Tail has 11–12 dark bands. Limbs usually darker than rest of body. Legs muscular with strongly curved claws, suitable for climbing and digging. DISTRIBUTION Found in the Philippines (Catanduanes, Luzon and Polillo). HABITS AND HABITAT Diurnal, arboreal and secretive. Feeds on fruits and invertebrates such as insects, crustaceans and snails. Female lays 4–16 eggs per clutch. Average hatchling 160mm SVL; 407mm TL. Needs intact forest to thrive.

Palawan Water Monitor ■ *Varanus palawanensis* SVL 790mm

DESCRIPTION Large monitor with robust body and limbs. Head, dorsal body, and tail dark grey to black, with or without yellow spots. Nostrils round to oblong near snout-tip. Throat off white or cream with dark marbling. Ventral body off white to yellowish. DISTRIBUTION Found in the Philippines (Palawan and surrounding islands). HABITS AND HABITAT Feeds on small vertebrates, invertebrates and carrion. Inhabits various ecosystems, such as lowland forests, mangroves, beaches and disturbed habitats.

Monitor Lizards

Emerald Tree Monitor — *Varanus prasinus* SVL 310mm

DESCRIPTION Overall bright green with 11–18 black bands from neck to dorsal body. Head, body and limbs slender. Soles of feet have enlarged dark scales. Toes long with sharp claws. Prehensile tail 1.5 times (adult) to two times (juvenile) SVL. **DISTRIBUTION** Found in Indonesia (Papua, West Papua and surrounding smaller islands). **HABITS AND HABITAT** Arboreal. Feeds on invertebrates such as cockroaches, stick insects and katydids. Inhabits lowland rainforests, palm forests and mangrove forests; also noted in plantations.

■ Monitor Lizards ■

Rough-necked Monitor
■ *Varanus rudicollis* SVL 590mm

DESCRIPTION Slender monitor with laterally flattened tail. Adults dark grey or black with 5–6 transverse rows of grey, brown or yellowish spots. Nuchal scales very enlarged. Nostrils oval (juveniles) to slit-shaped (adults), and positioned closer to eyes than to snout-tip. Juveniles have yellow spots on dorsal body and scattered bright yellow scales on limbs. **DISTRIBUTION** Found in Indonesia, Malaysia, Myanmar and Thailand. **HABITS AND HABITAT** Observations in the wild infrequent due to its shy nature. Apparently feeds on the ground and retreats to trees when threatened. Hatchling 20–26cm TL. Occurs in secondary and primary forests and mangrove swamps.

Monitor Lizards

Common Water Monitor ■ *Varanus salvator* SVL 1,170mm

DESCRIPTION Common large varanid. Dark/brown dorsal base colour (*V. s. macromaculatus*) with 4–7 rows of spots, some more or less distinctive. In between bands slightly marbled, with dark tail showing transverse rows of spots or fused spots resulting in cross-bands. Head dark; white chin with 1–3 lateral dark cross-bars; neck brown and may have light dots. A population of *V. s. macromaculatus* occurring in the Thai-Malaysia border area is completely black and commonly referred to as 'black dragon'. *V. s. bivittatus* only differs from *V. s. macromaculatus* by having nostril positions closer to snout-tip and shorter head; dorsal spots in *V. s. macromaculatus* tend to fuse to light cross-bands. **DISTRIBUTION** Two subspecies found in Southeast Asia: *V. s. macromaculatus* on Southeast Asian mainland, Indonesia (Sumatra), Borneo and several offshore islands; *V. s. bivittatus* in Indonesia, on Java, Bali, Lombok, Sumbabawa, Flores, Alor, Wetar and smaller offshore islands. **HABITS AND HABITAT** Semi-aquatic. Oviparous, laying multiple clutches a year of 5–40 eggs. Opportunistic and can inhabit various habitats, such as primary forests and mangrove swamps, but also places with human disturbance like agricultural areas and canal systems in cities. Mainly found in mangrove/swamp vegetation and wetlands below 1,800m asl, but usually below 1,000m asl.

Monitor Lizards

Timor Monitor *Varanus timorensis* SVL 250mm

DESCRIPTION Medium to dark brown dorsally and white or yellowish ventrally. Dorsal body has yellowish eye spots or reticulation. Tail circular, without keel. Spike-like scale on each side of cloaca. Juveniles have 12 rows of yellow spots on dorsal body, bright spots on head and legs, and narrow bright bands on tail. **DISTRIBUTION** Found in Indonesia (Timor) and Timor-Leste. **HABITS AND HABITAT** Feeds on invertebrates (insects, arachnids), and occasionally geckos and small snakes. Inhabits beach forests and disturbed areas such as rice fields and human settlements.

CHECKLIST OF THE LIZARDS OF SOUTHEAST ASIA

Global status according to the IUCN Red List of Threatened Species 2021 (version 2021–1).
Not all species have common names, and these are provided where available. Species in **bold** are described in this book.

Symbols:
X Present   ? Suspected to occur or doubtful records

Critically Endangered (CR) Species that meet any of the categories and corresponding criteria A–E for Critically Endangered and are considered to face an extremely high risk of extinction in the wild.

Endangered (EN) Species that meet any of the categories and corresponding criteria A–E for Endangered and are considered to face a very high risk of extinction in the wild.

Vulnerable (VU) Species that meet any of the categories and corresponding criteria A–E for Vulnerable and are considered to face a high risk of exctinction in the wild.

Near Threatened (NT) Species that do not qualify for the categories and corresponding criteria A–E for Critically Endangered, Endangered or Vulnerable, but are close to or expected to meet them in the near future.

Least Concern (LC) Species that do not qualify for the categories and corresponding criteria A–E for Critically Endangered, Enangered, Vulnerable or Near Threatened. This category often includes common and widespread species.

Data Deficient (DD) Species are considered Data Deficient when sufficient information is not available to make a direct or indirect assessment of their risk of extinction. More information is required for these taxa, in particular on aspects like abundance and range.

Common Name	Scientific name	Brunei Darussalam	Cambodia	Indonesia	Laos	Malaysia	Myanmar	The Philippines	Singapore	Thailand	Timor-Leste	Vietnam	Global Status
Agamidae (Agamids)													
Peninsular Horned Tree Lizard	*Acanthosaura armata*		x	x		x	x			x			
Orange-crested Horned Lizard	*Acanthosaura aurantiacrista*									x			
Bukit Larut Mountain Horned Agamid	*Acanthosaura bintangensis*					x							DD
	Acanthosaura brachypoda											x	DD
Green Pricklenape	*Acanthosaura capra*		x									x	NT
Cardamom Mountain Horned Agamid	*Acanthosaura cardamomensis*		x							x		x	LC
Coronated Agamid	*Acanthosaura coronata*		x									x	LC
Boulenger's Pricklenape	*Acanthosaura crucigera*				x	x	x			x		x	LC
Brown Pricklenape	*Acanthosaura lepidogaster*		x		x	x				x		x	LC
	Acanthosaura murphyi											x	
Natalia's Spiny Lizard	*Acanthosaura nataliae*				x							x	LC

Common Name	Scientific name	Brunei Darussalam	Cambodia	Indonesia	Laos	Malaysia	Myanmar	The Philippines	Singapore	Thailand	Timor-Leste	Vietnam	Global Status
Phong Điên Horned Agamid	Acanthosaura phongdienensis											x	
Phuket Horned Tree Agamid	Acanthosaura phuketensis									x			LC
	Acanthosaura prasina									x			
Titiwangsa Horned Tree Lizard	Acanthosaura titiwangsaensis					x							EN
Indonesia Earless Agama	Aphaniotis acutirostris			x									
Earless Agamid	Aphaniotis fusca			x		x			x	x			LC
Ornate Earless Agama	Aphaniotis ornata	x		x		x							LC
Burmese Green Crested Lizard	Bronchocela burmana						x						LC
Sulawesi Bloodsucker	Bronchocela celebensis			x									
Green Crested Lizard	Bronchocela cristatella	x		x		x	x	x	x	x			
Sumatra Bloodsucker	Bronchocela hayeki			x									
Maned Forest Lizard	Bronchocela jubata		x	x					x		x		LC
Marbled Green Crested Lizard	Bronchocela marmorata							x					DD
Orlov's Forest Lizard	Bronchocela orlovi											x	DD
Gunung Raya Green-crested Lizard	Bronchocela rayaensis					x							LC
	Bronchocela shenlong											x	NT
Günther's Bloodsucker	Bronchocela smaragdina		x							x		x	LC
	Bronchocela vietnamensis											x	VU
Vietnamese Blue Crested Lizard	Calotes bachae											x	LC
	Calotes chincollium						x						LC
Forest Crested Lizard	Calotes emma		x		x	x	x			x		x	
	Calotes htunwini						x						LC
	Calotes irawadi						x						LC
Jerdon's Forest Lizard	Calotes jerdoni						x						
Blue Crested Lizard	Calotes mystaceus	x	x		x					x		x	
	Calotes nigriplicatus				x								
Oriental Garden Lizard	Calotes versicolor	x	x	x	x	x	x	x	x	x		x	
Blackthroated Bloodsucker	Compliticus nigrigularis						x						DD
Smooth-scaled Mountain Lizard	Cristidorsa planidorsata						x						
	Dendragama australis			x									
Boulenger's Tree Agama	Dendragama boulengeri			x									
	Dendragama dioidema			x									
Schneider's Tree Agama	Dendragama schneideri			x									
Japalure	Diploderma chapaense											x	DD
Banded Japalure	Diploderma fasciatum											x	LC
Hampton's Japalure	Diploderma hamptoni						x						DD
	Diploderma ngoclinense											x	
Yunnan Japalure	Diploderma yunnanensis						x						
Singapore Flying Dragon	Draco abbreviatus			x					x				
	Draco beccarii			x									
Lazell's Flying Dragon	Draco biaro			x									
Two-spotted Flying Lizard	Draco bimaculatus						x						LC
Blanford's Flying Lizard	Draco blanfordii				x	x				x		x	LC
Boschma's Gliding Lizard	Draco boschmai			x									
Sangihe Flying Dragon	Draco caerulhians			x									
Horned Flying Lizard	Draco cornutus	x		x	x								LC

148

Common Name	Scientific name	Brunei Darussalam	Cambodia	Indonesia	Laos	Malaysia	Myanmar	The Philippines	Singapore	Thailand	Timor-Leste	Vietnam	Global Status
Crested Flying Dragon	*Draco cristatellus*					x							DD
Chartreuse-spotted Flying Lizard	*Draco cyanopterus*							x					LC
Fringed Flying Dragon	*Draco fimbriatus*			x		x		x		x			
Dusky Gliding Lizard	*Draco formosus*					x				x			LC
Günther's Flying Lizard	*Draco guentheri*						x						LC
Red-beared Flying Lizard	*Draco haematopogon*			x		x				x			LC
Indochinese Flying Lizard	*Draco indochinensis*		x									x	LC
Iskandar's Flying Lizard	*Draco iskandari*			x									
Batan Flying Lizard	*Draco jareckii*							x					LC
Lined Flying Dragon	*Draco lineatus*			x		x		x					LC
Spotted Flying Lizard	*Draco maculatus*		x		x	x	x			x		x	LC
Giant Gliding Lizard	*Draco maximus*			x		x				x			LC
Black-bearded Flying Lizard	*Draco melanopogon*	x		x		x			x	x			
Mindanao Gliding Lizard	*Draco mindanensis*							x					VU
Lined Flying Dragon	*Draco modiglianii*			x									
Dusky Gliding Lizard	*Draco obscurus*			x						x			LC
White-spotted Flying Lizard	*Draco ornatus*							x					LC
Palawan Gliding Lizard	*Draco palawanensis*							x					LC
	Draco punctatus									x			
Quadras's Flying Lizard	*Draco quadrasi*							x					LC
Five-lined Flying Lizard	*Draco quinquefasciatus*	x		x		x				x	x		
Reticulated Flying Lizard	*Draco reticulatus*							x					LC
	Draco rhytisma			x									
Sulawesi Lined Gliding Lizard	*Draco spilonotus*			x									
Philippine Flying Dragon	*Draco spilopterus*			x				x					LC
Sumatran Gliding Lizard	*Draco sumatranus*	x				x			x	x			LC
	Draco supriatnai			x									
Barred Gliding Lizard	*Draco taeniopterus*		x			x	x			x		x	LC
Timor Flying Dragon	*Draco timoriensis*			x							x		
Common Gliding Lizard	*Draco volans*			x									LC
	Draco walkeri			x									
Abbot's Anglehead Lizard	*Gonocephalus abbotti*					x				x			LC
Bell's Forest Dragon	*Gonocephalus bellii*			x		x				x			LC
Sumatra Forest Dragon	*Gonocephalus beyschlagi*			x									
Borneo Forest Dragon	*Gonocephalus bornensis*	x		x		x							LC
Chameleon Forest Dragon	*Gonocephalus chamaeleontinus*			x		x							
Peter's Forest Dragon	*Gonocephalus doriae*			x		x				x			LC
Great Anglehead Lizard	*Gonocephalus grandis*	x		x	x	x				x		x	LC
Boulenger's Forest Dragon	*Gonocephalus interruptus*							x					DD
Kloss' Forest Dragon	*Gonocephalus klossi*			x									
Kuhl's Anglehead Lizard	*Gonocephalus kuhlii*			x									
Manthey's Forest Dragon	*Gonocephalus lacunosus*			x									DD
Blue-eyed Anglehead Lizard	*Gonocephalus liogaster*	x		x		x				x			
Bleeker's Forest Dragon	*Gonocephalus megalepis*			x									
	Gonocephalus mjobergi			x		x							DD
Mindoro Forest Dragon	*Gonocephalus semperi*							x					DD

Common Name	Scientific name	Brunei Darussalam	Cambodia	Indonesia	Laos	Malaysia	Myanmar	The Philippines	Singapore	Thailand	Timor-Leste	Vietnam	Global Status
Phillippine Forest Dragon	Gonocephalus sophiae							x					DD
Sumatra Nose-horned Lizard	Harpesaurus beccarii			x									
Bornean Nose-horned Lizard	Harpesaurus borneensis			x		x							DD
	Harpesaurus brooksi			x									
Nias Nose-horned Lizard	Harpesaurus ensicauda			x									
Modigliani's Nose-horned Lizard	Harpesaurus modiglianii			x									DD
Java Nose-horned Lizard	Harpesaurus tricinctus			x		x							
Amboina Sailfin Lizard	Hydrosaurus amboinensis			x									
Sulawesi Sailfin Lizard	Hydrosaurus celebensis			x									
	Hydrosaurus microlophus			x									
Philippine Sailfin Lizard	Hydrosaurus pustulatus							x					VU
Weber's Sailfin Lizard	Hydrosaurus weberi			x									
Kinabalu Crested Dragon	Hypsicalotes kinabaluensis			x		x							DD
	Hypsilurus auritus			x									LC
Two-marked Forest Dragon	Hypsilurus binotatus			x									LC
Bruijni Forest Dragon	Hypsilurus bruijnii			x									DD
New Guinea Forest Dragon	Hypsilurus geelvinkianus			x									LC
Hikida's Forest Dragon	Hypsilurus hikidanus			x									LC
	Hypsilurus magnus			x									LC
Modest Forest Dragon	Hypsilurus modestus			x									LC
	Hypsilurus nigrigularis			x									DD
	Hypsilurus schultzewestrumi			x									LC
	Hypsilurus spinosus			x									
	Hypsilurus tenuicephalus			x									DD
Burmese Japalure	Japalura sagittifera						x						
Beauty Butterfly Lizard	Leiolepis belliana	x	x	x	x	x				x		x	LC
Böhme's Butterfly Lizard	Leiolepis boehmei									x			VU
Peters' Butterfly Lizard	Leiolepis guentherpetersi											x	EN
Spotted Butterfly Lizard	Leiolepis guttata											x	DD
Ngo Van Tri's Lady Butterfly Lizard	Leiolepis ngovantrii											x	VU
Burmese Butterfly Lizard	Leiolepis peguensis						x						LC
Reeves' Butterfly Lizard	Leiolepis reevesii	x								x		x	
Red-banded Butterfly Lizard	Leiolepis rubritaeniata	x	x							x		x	LC
Malayan Butterfly Lizard	Leiolepis triploida					x				x			DD
	Lophocalotes achlios			x									
Crested Lizard	Lophocalotes ludekingi			x									
	Lophognathus maculilabris			x									
Indonesian Forest Dragon	Lophosaurus dilophus			x									LC
Robinson's Anglehead Lizard	Malayodracon robinsonii					x							NT
Phu Wua Lizard	Mantheyus phuwuanensis				x					x			NT
Sabah Eyebrow Lizard	Pelturagonia borneensis			x									
Mocquard's Eyebrow Lizard	Pelturagonia cephalum			x									LC
Black-lipped Shrub Lizard	Pelturagonia nigrilabris			x		x							
Sarawak Eyebrow Lizard	Pelturagonia spiniceps			x		x							
Hubrecht's Eyebrow Lizard	Phoxophrys tuberculata			x									
Chinese Water Dragon	Physignathus cocincinus		x	x						x		x	VU

Common Name	Scientific name	Brunei Darussalam	Cambodia	Indonesia	Laos	Malaysia	Myanmar	The Philippines	Singapore	Thailand	Timor-Leste	Vietnam	Global Status
	Pseudocalotes baliomus			x									
Vietnam False Bloodsucker	Pseudocalotes brevipes											x	LC
	Pseudocalotes cybelidermus			x									
	Pseudocalotes dringi					x							LC
Drogon's False Garden Lizard	Pseudocalotes drogon					x							DD
Yellow-throated False Garden Lizard	Pseudocalotes flavigula					x							CR
Flower's Forest Agamid	Pseudocalotes floweri		x							x			VU
	Pseudocalotes guttalineatus			x									
Burmese Mountain Agamid	Pseudocalotes kakhienensis						x			x			
Khao Nan Long-headed Lizard	Pseudocalotes khaonanensis									x			DD
Kingdonward's Bloodsucker	Pseudocalotes kingdonwardi						x						LC
Bukit Larut False Garden Lizard	Pseudocalotes larutensis					x							VU
Small-scaled Montane Forest Lizard	Pseudocalotes microlepis				x	x				x		x	
Laotian False Bloodsucker	Pseudocalotes poilani				x								EN
Rhaegal's False Garden Lizard	Pseudocalotes rhaegal					x							CR
	Pseudocalotes rhammanotus			x									
	Pseudocalotes saravacensis					x							DD
Indonesian False Bloodsucker	Pseudocalotes tympanistriga			x									
Viserion's False Garden Lizard	Pseudocalotes viserion					x							VU
Ziegler's Tree Lizard	Pseudocalotes ziegleri											x	DD
	Pseudocophotis kontumensis											x	DD
	Pseudocophotis sumatrana			x									DD
	Ptyctolaemus collicristatus							x					LC
Green Fan-throated Lizard	Ptyctolaemus gularis							x					
Northern Water Dragon	Tropicagama temporalis			x									LC
Anguidae (Glass Lizards)													
Buettikofer's Glass Lizard	Dopasia buettikoferi			x		x							LC
Burmese Glass Lizard	Dopasia gracilis						x			x		x	
Hart's Glass Lizard	Dopasia harti											x	LC
Ludovic's Glass Lizard	Dopasia ludovici											x	LC
Buon Luoi Glass Lizard	Dopasia sokolovi											x	LC
Wegner's Glass Lizard	Dopasia wegneri			x									DD
Dibamidae (Blind Lizards)													
Alfred's Worm Lizard	Dibamus alfredi			x		x				x			DD
Boo Liat's Worm Lizard	Dibamus booliati					x							DD
White-tailed Dibamid	Dibamus bourreti											x	LC
	Dibamus celebensis			x									
	Dibamus dalaiensis		x										LC
	Dibamus deharvengi											x	DD
	Dibamus dezwaani			x									
Flower's Blind Lizard	Dibamus floweri									x			
Greer's Blind Skink	Dibamus greeri											x	LC
	Dibamus ingeri					x	x						DD
	Dibamus kondaoensis											x	NT
White-tailed Worm Lizard	Dibamus leucurus			x		x		x					LC
Mountain Blind Skink	Dibamus montanus											x	DD

Common Name	Scientific name	Brunei Darussalam	Cambodia	Indonesia	Laos	Malaysia	Myanmar	The Philippines	Singapore	Thailand	Timor-Leste	Vietnam	Global Status
New Guinea Blind Earless Skink	*Dibamus novaeguineae*	x		x				x					LC
Seram Blind Skink	*Dibamus seramensis*			x									
Smith's Blind Skink	*Dibamus smithi*									x			DD
Somsak's Dibamid Lizard	*Dibamus somsaki*									x			DD
Taylor's Blind Skink	*Dibamus taylori*			x									
	Dibamus tebal			x									
Tioman Island Blind Lizard	*Dibamus tiomanensis*					x							EN
	Dibamus vorisi			x		x							DD
Eublepharidae (Eyelid Geckos)													
Cat Gecko	*Aeluroscalabotes felinus*			x		x				x		x	LC
Vietnamese Leopard Gecko	*Goniurosaurus araneus*											x	
Cat Ba Leopard Gecko	*Goniurosaurus catbaensis*											x	EN
Huu Lien Leopard Gecko	*Goniurosaurus huuliensis*											x	CR
Lichtenfelder's Gecko	*Goniurosaurus lichtenfelderi*											x	VU
Chinese Leopard Gecko	*Goniurosaurus luii*											x	
Gekkonidae (Typical Geckos)													
Aceh Round-eyed Gecko	*Cnemaspis aceh*			x									
	Cnemaspis adangrawi									x			
Penang Island Round-eyed Gecko	*Cnemaspis affinis*					x				x			VU
	Cnemaspis andalas			x									
Mount Lawit Round-eyed Gecko	*Cnemaspis argus*					x							LC
Saffron-limbed Round-eyed Gecko	*Cnemaspis aurantiacopes*									x			DD
Bauer's Rock Gecko	*Cnemaspis baueri*					x							LC
Bayu Cave Round-eyed Gecko	*Cnemaspis bayuensis*					x							VU
Pulau Bidong Round-eyed Gecko	*Cnemaspis bidongensis*					x							LC
Twin-spotted Rock Gecko	*Cnemaspis biocellata*					x				x			LC
Con Dao Round-eyed Gecko	*Cnemaspis boulengeri*											x	NT
White-tailed Round-eyed Gecko	*Cnemaspis caudanivea*											x	VU
Chan-ard's Rock Gecko	*Cnemaspis chanardi*					x				x			LC
Chanthaburi Rock Gecko	*Cnemaspis chanthaburiensis*		x							x			LC
Nias Round-eyed Gecko	*Cnemaspis dezwaani*			x									
Sarawak Round-eyed Gecko	*Cnemaspis dringi*			x		x							DD
Orange-bellied Round-eyed Gecko	*Cnemaspis flavigaster*					x							LC
Titiwangsa Round-eyed Gecko	*Cnemaspis flavolineata*					x				x			DD
Grismer's Round-eyed Gecko	*Cnemaspis grismeri*					x							LC
Bukit Hangus Round-eyed Gecko	*Cnemaspis hangus*					x							NT
Tiger Round-eyed Gecko	*Cnemaspis harimau*					x							VU
Orange-headed Round-eyed Gecko	*Cnemaspis huaseesom*									x			LC
Simeulue Round-eyed Gecko	*Cnemaspis jacobsoni*			x									DD
Kamolnorranath's Round-eyed Gecko	*Cnemaspis kamolnorranathi*									x			LC
Karst-dwelling Round-eyed Gecko	*Cnemaspis karsticola*					x							VU
Kendall's Rock Gecko	*Cnemaspis kendallii*			x		x			x				LC
Kumpol's Rock Gecko	*Cnemaspis kumpoli*					x				x			LC
Laotian Round-eyed Gecko	*Cnemaspis laoensis*				x								DD
Curse Round-eyed Gecko	*Cnemaspis leucura*					x							DD
Tioman Island Rock Gecko	*Cnemaspis limi*					x							LC

Common Name	Scientific name	Brunei Darussalam	Cambodia	Indonesia	Laos	Malaysia	Myanmar	The Philippines	Singapore	Thailand	Timor-Leste	Vietnam	Global Status
Lan Saka Rock Gecko	Cnemaspis lineatubercularis									x			
Stripe-throated Round-eyed Gecko	Cnemaspis lineogularis									x			LC
Mahsuri's Round-eyed Gecko	Cnemaspis mahsuriae					x							LC
McGuire's Rock Gecko	Cnemaspis mcguirei					x				x			LC
	Cnemaspis minang			x									
Enggano Round-eyed Gecko	Cnemaspis modiglianii			x									
Wanaram Temple Round-eyed Gecko	Cnemaspis monachorum					x							LC
Mumpuni Round-eyed Gecko	Cnemaspis mumpuniae			x									
	Cnemaspis muria			x									
Narathiwat Round-eyed Gecko	Cnemaspis narathiwatensis									x			LC
Cambodian Round-eyed Gecko	Cnemaspis neangthyi		x										EN
Black-spotted Round-eyed Gecko	Cnemaspis nigridia					x	x						LC
Palean Round-eyed Gecko	Cnemaspis niyomwanae					x				x			EN
Nuicam Round-eyed Gecko	Cnemaspis nuicamensis											x	VU
Omar's Round-eyed Gecko	Cnemaspis omari					x							LC
South Pagai Round-eyed Gecko	Cnemaspis pagai			x									
Fairy Rock Gecko	Cnemaspis paripari					x	x						CR
Pemanggil Island Round-eyed Gecko	Cnemaspis pemanggilensis					x							LC
Peninsular Rock Gecko	Cnemaspis peninsularis					x			x				LC
Perhentian Islands Round-eyed Gecko	Cnemaspis perhentianensis					x							NT
Phang Nga Round-eyed Gecko	Cnemaspis phangngaensis									x			LC
Phuket Rock Gecko	Cnemaspis phuketensis									x			LC
Larut Hills Round-eyed Gecko	Cnemaspis pseudomcguirei					x							NT
Psychedelic Rock Gecko	Cnemaspis psychedelica											x	EN
Spotted-neck Round-eyed Gecko	Cnemaspis punctatonuchalis									x			LC
Belitung Island Round-eyed Gecko	Cnemaspis purnamai			x									
Mount Rajabasa Round-eyed Gecko	Cnemaspis rajabasa			x									
Roti Canai Round-eyed Gecko	Cnemaspis roticanai					x							LC
Merapoh Round-eyed Gecko	Cnemaspis selamatkanmerapoh					x							VU
Moon Rabbit Rock Gecko	Cnemaspis selenolagus											x	
Telok Bahang Round-eyed Gecko	Cnemaspis shahruli					x							LC
Siamese Round-eyed Gecko	Cnemaspis siamensis									x			LC
Gunung Stong Round-eyed Gecko	Cnemaspis stongensis					x							LC
Anambas Round-eyed Gecko	Cnemaspis sundagekko			x									
	Cnemaspis sundainsula			x									
	Cnemaspis tanintharyi						x						
	Cnemaspis tapanuli			x									
	Cnemaspis tarutaoensis									x			
Temiah Round-eyed Gecko	Cnemaspis temiah					x							CR
Tha Chana Round-eyed Gecko	Cnemaspis thachanaensis									x			NT
	Cnemaspis thayawthadangyi						x						
	Cnemaspis tubaensis					x							
	Cnemaspis tucdupensis											x	VU
Van Deventer's Rock Gecko	Cnemaspis vandeventeri											x	LC
Siberut Round-eyed Gecko	Cnemaspis whittenorum			x									
Wondiwoi Bent-toed Gecko	Cyrtodactylus aaroni			x									DD

Common Name	Scientific name	Brunei Darussalam	Cambodia	Indonesia	Laos	Malaysia	Myanmar	The Philippines	Singapore	Thailand	Timor-Leste	Vietnam	Global Status
Mon State Bent-toed Gecko	Cyrtodactylus aequalis						x						DD
Agam Bent-toed Gecko	Cyrtodactylus agamensis			x									
Agusan Bent-toed Gecko	Cyrtodactylus agusanensis							x					LC
	Cyrtodactylus amphipetraeus									x			
Angled Forest Gecko	Cyrtodactylus angularis									x			LC
Sagaing Bent-toed Gecko	Cyrtodactylus annandalei						x						LC
Annulated Bent-toed Gecko	Cyrtodactylus annulatus							x					LC
Stardust Bent-toed Gecko	Cyrtodactylus astrum					x				x			LC
Kyauk Nagar Cave Bent-toed Gecko	Cyrtodactylus aunglini						x						
Phnom Aural Bent-toed Gecko	Cyrtodactylus auralensis		x										
Aur Island Bent-toed Gecko	Cyrtodactylus aurensis					x							LC
Golden-belted Bent-toed Gecko	Cyrtodactylus auribalteatus									x			VU
Southern Titiwangsa Bent-toed Gecko	Cyrtodactylus australotitiwangsaensis					x							LC
Ayeyarwady Bent-toed Gecko	Cyrtodactylus ayeyarwadyensis						x						DD
Ba Den Bent-toed Gecko	Cyrtodactylus badenensis											x	VU
Balu Bent-toed Gecko	Cyrtodactylus baluensis			x		x							LC
Ban Soc Bent-toed Gecko	Cyrtodactylus bansocensis				x								DD
Batik Bent-toed Gecko	Cyrtodactylus batik			x									
Besar Island Bent-toed Gecko	Cyrtodactylus batucolus					x							LC
Bayin Nyi Cave Bent-toed Gecko	Cyrtodactylus bayinnyiensis						x						
Bich Ngan's Bent-toed Gecko	Cyrtodactylus bichnganae											x	VU
Mount Bidoup Bent-toed Gecko	Cyrtodactylus bidoupimontis											x	LC
Bintang Lowland Bent-toed Gecko	Cyrtodactylus bintangrendah					x							LC
Bintang Mountain Bent-toed Gecko	Cyrtodactylus bintangtinggi					x				x			NT
Bobrov's Bent-toed Gecko	Cyrtodactylus bobrovi											x	LC
Bokor Bent-toed Gecko	Cyrtodactylus bokorensis		x										
Short-toed Bent-toed Gecko	Cyrtodactylus brevidactylus						x						EN
Short-handed Bent-toed Gecko	Cyrtodactylus brevipalmatus									x			LC
Laotian Bent-toed Gecko	Cyrtodactylus buchardi				x								DD
Bu Gia Map Bent-toed Gecko	Cyrtodactylus bugiamapensis											x	LC
Calame's Bent-toed Gecko	Cyrtodactylus calamei				x								LC
Ninh Thuan Bent-toed Gecko	Cyrtodactylus caovansungi											x	EN
Cardamom Mountains Bent-toed Gecko	Cyrtodactylus cardamomensis		x										
Cattien Bent-toed Gecko	Cyrtodactylus cattienensis											x	LC
Sarawak Bent-toed Gecko	Cyrtodactylus cavernicolus					x							VU
	Cyrtodactylus celatus			x							x		
Saraburi Bent-toed Gecko	Cyrtodactylus chanhomeae									x			CR
Chaunghanakwa Hill Bent-toed Gecko	Cyrtodactylus chaunghanakwaensis						x						
Chauquang Bent-toed Gecko	Cyrtodactylus chauquangensis											x	LC
Shan State Bent-toed Gecko	Cyrtodactylus chrysopylos						x						VU
Pulo Condore Bow-fingered Gecko	Cyrtodactylus condorensis											x	LC
Dawei Bent-toed Gecko	Cyrtodactylus consobrinoides						x						DD
Banded Bent-toed Gecko	Cyrtodactylus consobrinus	x	x	x					x	x			
Cryptic Bent-toed Gecko	Cyrtodactylus cryptus				x							x	LC
Cucdong Bent-toed Gecko	Cyrtodactylus cucdongensis											x	LC

Common Name	Scientific name	Brunei Darussalam	Cambodia	Indonesia	Laos	Malaysia	Myanmar	The Philippines	Singapore	Thailand	Timor-Leste	Vietnam	Global Status
Cuc Phuong Bent-toed Gecko	Cyrtodactylus cucphuongensis											x	LC
	Cyrtodactylus culaochamensis											x	
Dammathet Cave Bent-toed Gecko	Cyrtodactylus dammathetensis						x						
Darevsky's Bent-toed Gecko	Cyrtodactylus darevskii				x								DD
Darmandville Bow-fingered Gecko	Cyrtodactylus darmandvillei			x									
Bu Dop Bent-toed Gecko	Cyrtodactylus dati		x									x	DD
	Cyrtodactylus dattkyaikensis						x						
	Cyrtodactylus dayangbuntingensis					x							
Moluccan Bow-fingered Gecko	Cyrtodactylus deveti			x									DD
Doi Suthep Bent-toed Gecko	Cyrtodactylus doisuthep									x			LC
Dumnui's Bent Toed Gecko	Cyrtodactylus dumnuii									x			LC
Durian Bent-toed Gecko	Cyrtodactylus durio					x							DD
Eisenman's Bent-toed Gecko	Cyrtodactylus eisenmanae											x	LC
White-eyed Bent-toed Gecko	Cyrtodactylus elok					x				x			LC
Red-eyed Bent-toed Gecko	Cyrtodactylus erythrops									x			LC
	Cyrtodactylus evanquahi					x							
Puepoli Bent-toed Gecko	Cyrtodactylus feae						x			x			DD
Tamarind Bent-toed Gecko	Cyrtodactylus fumosus				x								
Min Dat Bent-toed Gecko	Cyrtodactylus gansi						x						LC
Gialai Bent-toed Gecko	Cyrtodactylus gialaiensis											x	CR
Lombok Bent-toed Gecko	Cyrtodactylus gordongekkoi			x									DD
Grismer's Bent-toed Gecko	Cyrtodactylus grismeri											x	VU
Gua Kanthan Bent-toed Gecko	Cyrtodactylus guakanthanensis					x							CR
Leyte Bent-toed Gecko	Cyrtodactylus gubaot							x					
	Cyrtodactylus gunungsenyumensis					x							LC
Halmahera Bent-toed Gecko	Cyrtodactylus halmahericus			x									
Chiku Bent-toed Gecko	Cyrtodactylus hidupselamanya					x							VU
Bunguran Bent-toed Gecko	Cyrtodactylus hikidai			x									
Hinnamno Bent-toed Gecko	Cyrtodactylus hinnamnoensis				x								LC
Mekongga Mountains Bent-toed Gecko	Cyrtodactylus hitchi			x									
Hon Tre Bent-toed Gecko	Cyrtodactylus hontreensis											x	LC
	Cyrtodactylus houaphanensis				x							x	
Chua Chan Bent-toed Gecko	Cyrtodactylus huynhi											x	VU
Sabah Bow-fingered Gecko	Cyrtodactylus ingeri	x		x									LC
Phetchabun Bent-toed Gecko	Cyrtodactylus interdigitalis				x					x			LC
Intermediate Bent-toed Gecko	Cyrtodactylus intermedius									x			LC
Doi Inthanon Bent-toed Gecko	Cyrtodactylus inthanon									x			LC
West Irian Bent-toed Gecko	Cyrtodactylus irianjayaensis			x									DD
Irregular Bow-fingered Gecko	Cyrtodactylus irregularis											x	DD
Jaeger's Bent-toed Gecko	Cyrtodactylus jaegeri				x								CR
Jambangan Bent-toed Gecko	Cyrtodactylus jambangan							x					
Jarak Island Bent-toed Gecko	Cyrtodactylus jarakensis					x							CR
Jarujin's Bent-toed Gecko	Cyrtodactylus jarujini				x					x			LC
	Cyrtodactylus jatnai			x									

Common Name	Scientific name	Brunei Darussalam	Cambodia	Indonesia	Laos	Malaysia	Myanmar	The Philippines	Singapore	Thailand	Timor-Leste	Vietnam	Global Status
Jelawang Bent-toed Gecko	Cyrtodactylus jelawangensis					x							DD
Kabaena Bow-fingered Gecko	Cyrtodactylus jellesmae			x									
Khammouane Bent-toed Gecko	Cyrtodactylus khammouanensis				x								VU
Lampang Bent-toed Gecko	Cyrtodactylus khelangensis									x			EN
Kingsada's Bent-toed Gecko	Cyrtodactylus kingsadai											x	LC
	Cyrtodactylus kohrongensis		x										
Kunya's Bent-toed Gecko	Cyrtodactylus kunyai									x			LC
Phnom Laang Bent-toed Gecko	Cyrtodactylus laangensis		x										
Komodo Bent-toed Gecko	Cyrtodactylus laevigatus			x									
Langkawi Island Bent-toed Gecko	Cyrtodactylus langkawiensis					x							LC
Spiny Forest Gecko	Cyrtodactylus lateralis			x		x							
Tenggol Island Bent-toed Gecko	Cyrtodactylus leegrismeri					x							LC
Lekagul's Bent-toed Gecko	Cyrtodactylus lekaguli									x			LC
Lenggong Bent-toed Gecko	Cyrtodactylus lenggongensis					x							LC
Lenya Banded Bent-toed Gecko	Cyrtodactylus lenya						x						DD
Five-banded Bent-toed Gecko	Cyrtodactylus limajur					x							
Linno Cave Bent-toed Gecko	Cyrtodactylus linnoensis						x						
Linn-Way Bent-toed Gecko	Cyrtodactylus linnwayensis						x						
Lomyen Bent-toed Gecko	Cyrtodactylus lomyenensis				x								VU
Large-tubercled Bent-toed Gecko	Cyrtodactylus macrotuberculatus					x							LC
	Cyrtodactylus maelanoi									x			
Singapore Bent-toed Gecko	Cyrtodactylus majulah			x					x				
Borneo Bow-fingered Gecko	Cyrtodactylus malayanus	x		x		x							LC
Dinagat Bent-toed Gecko	Cyrtodactylus mamanwa							x					
Mandalay Bent-toed Gecko	Cyrtodactylus mandalayensis						x						DD
Marbled Bent-toed Gecko	Cyrtodactylus marmoratus			x		x							LC
Martin's Bent–toed Gecko	Cyrtodactylus martini											x	DD
Hikida's Bow-fingered Gecko	Cyrtodactylus matsuii					x							LC
Bago Yoma Bent-toed Gecko	Cyrtodactylus meersi						x						
Batu Caves Bent-toed Gecko	Cyrtodactylus metropolis					x							LC
False Bow-fingered Gecko	Cyrtodactylus mimikanus			x									LC
	Cyrtodactylus mombergi						x						
	Cyrtodactylus muangfuangensis				x								
Multipored Bent-toed Gecko	Cyrtodactylus multiporus				x								LC
Mulu Bent-toed Gecko	Cyrtodactylus muluensis					x							
Mya Leik Taung Bent-toed Gecko	Cyrtodactylus myaleiktaung						x						
Mount Popa Bent-toed Gecko	Cyrtodactylus myintkyawthurai						x						
Naung Ka Yaing Bent-toed Gecko	Cyrtodactylus naungkayaingensis						x						
	Cyrtodactylus ngoiensis				x								
Black-eyed Bent-toed Gecko	Cyrtodactylus nigriocularis											x	CR
Seram Bent-toed Gecko	Cyrtodactylus nuaulu			x									
Shwe Settaw Bent-toed Gecko	Cyrtodactylus nyinyikyawi						x						
Oldham's Bent-toed Gecko	Cyrtodactylus oldhami						x			x			LC
Ota's Bent-toed Gecko	Cyrtodactylus otai											x	EN

Common Name	Scientific name	Brunei Darussalam	Cambodia	Indonesia	Laos	Malaysia	Myanmar	The Philippines	Singapore	Thailand	Timor-Leste	Vietnam	Global Status
Vang Vieng Bent-toed Gecko	*Cyrtodactylus pageli*				x								LC
Panti Mountain Bent-toed Gecko	*Cyrtodactylus pantiensis*					x							VU
Butterfly Bent-toed Gecko	*Cyrtodactylus papilionoides*									x			LC
Papua Bow-fingered Gecko	*Cyrtodactylus papuensis*			x									LC
Paradox Bent-toed Gecko	*Cyrtodactylus paradoxus*											x	
Malaysian Swamp Bent-toed Gecko	*Cyrtodactylus payacola*					x							DD
Tenasserim Mountain Bent-toed Gecko	*Cyrtodactylus payarhtanensis*						x						DD
Pegu Bent-toed Gecko	*Cyrtodactylus peguensis*					x	x			x			LC
Farmer's Bent-toed Gecko	*Cyrtodactylus petani*			x									
Pharbaung Cave Bent-toed Gecko	*Cyrtodactylus pharbaungensis*						x						
Phetchaburi Bent-toed Gecko	*Cyrtodactylus phetchaburiensis*									x			LC
Philippine Bent-toed Gecko	*Cyrtodactylus philippinicus*					x		x	x				LC
	Cyrtodactylus phnomchiensis		x										
Phongnhakebang Bent-toed Gecko	*Cyrtodactylus phongnhakebangensis*											x	LC
Phuket Bent-toed Gecko	*Cyrtodactylus phuketensis*									x			EN
	Cyrtodactylus phumyensis											x	
Phước Bình Bent-toed Gecko	*Cyrtodactylus phuocbinhensis*											x	LC
Phu Quoc Bent–toed Gecko	*Cyrtodactylus phuquocensis*											x	EN
	Cyrtodactylus pinlaungensis						x						
Speckle-faced Bent-toed Gecko	*Cyrtodactylus psarops*		x										
Thua Thien-Hue Bent-toed Gecko	*Cyrtodactylus pseudoquadrivirgatus*											x	LC
Inger's Bent-toed Gecko	*Cyrtodactylus pubisulcus*	x				x							LC
Pù Hu Bent-toed Gecko	*Cyrtodactylus puhuensis*											x	LC
Malayan Bent-toed Gecko	*Cyrtodactylus pulchellus*					x							EN
Pyadalin Cave Bent-toed Gecko	*Cyrtodactylus pyadalinensis*						x						
Pyinyaung Bent-toed Gecko	*Cyrtodactylus pyinyaungensis*						x						
Four-striped Bent-toed Gecko	*Cyrtodactylus quadrivirgatus*		x			x			x	x			LC
Ranong Bent-toed Gecko	*Cyrtodactylus ranongensis*									x			LC
Palawan Bent-toed Gecko	*Cyrtodactylus redimiculus*							x					NT
Roesler's Bent-toed Gecko	*Cyrtodactylus roesleri*				x							x	LC
Rosichonarief's Bent-toed Gecko	*Cyrtodactylus rosichonarieforum*			x									
Andaman Bent-toed Gecko	*Cyrtodactylus rubidus*						x						
Rufford Bent-toed Gecko	*Cyrtodactylus rufford*		x										DD
Htamanthi Bent-toed Gecko	*Cyrtodactylus russelli*						x						DD
Sadan Cave Bent-toed Gecko	*Cyrtodactylus sadanensis*						x						
Sadan Sin Cave Bent-toed Gecko	*Cyrtodactylus sadansinensis*						x						
Sai Yok Bent-toed Gecko	*Cyrtodactylus saiyok*									x			LC
Sam Roi Yot Bent-toed Gecko	*Cyrtodactylus samroiyot*									x			LC
Nui Chua Bent-toed Gecko	*Cyrtodactylus sangi*											x	
Sanook Bent-toed Gecko	*Cyrtodactylus sanook*									x			LC
Sanpel Cave Bent-toed Gecko	*Cyrtodactylus sanpelensis*						x						
Johor Bent-toed Gecko	*Cyrtodactylus semenanjungensis*					x			x				NT
Semiadil's Bent-toed Gecko	*Cyrtodactylus semiadii*			x									
Kerinci Bent-toed Gecko	*Cyrtodactylus semicinctus*			x									

Common Name	Scientific name	Brunei Darussalam	Cambodia	Indonesia	Laos	Malaysia	Myanmar	The Philippines	Singapore	Thailand	Timor-Leste	Vietnam	Global Status
Bay Núi Bent-toed Gecko	Cyrtodactylus septimontium											x	
Seribuat Islands Bent-toed Gecko	Cyrtodactylus seribuatensis					x							VU
Sharkar's Bent-toed Gecko	Cyrtodactylus sharkari						x						VU
Shwetaung Bent-toed Gecko	Cyrtodactylus shwetaungorum						x						
Sin Yine Cave Bent-toed Gecko	Cyrtodactylus sinyineensis						x						
Monywa Bent-toed Gecko	Cyrtodactylus slowinskii						x						LC
Sommerlad's Bent-toed Gecko	Cyrtodactylus sommerladi			x									LC
	Cyrtodactylus soni											x	LC
	Cyrtodactylus sonlaensis											x	
Soudthichak's Bent-toed Gecko	Cyrtodactylus soudthichaki				x								LC
	Cyrtodactylus spelaeus				x								DD
Spiny Bent-toed Gecko	Cyrtodactylus spinosus			x									
Batang Padan Bent-toed Gecko	Cyrtodactylus stresemanni			x									DD
Sumontha's Bent-toed Gecko	Cyrtodactylus sumonthai									x			LC
Taft Forest Bent-toed Gecko	Cyrtodactylus sumuroi							x					
Surin Islands Bent-toed Gecko	Cyrtodactylus surin									x			LC
Kota Tinggi Bent-toed Gecko	Cyrtodactylus sworderi					x							EN
Sangir Island Bent-toed Gecko	Cyrtodactylus tahuna			x									
Ta Kou Bent-toed Gecko	Cyrtodactylus takouensis											x	CR
Nam Tamai Valley Bent-toed Gecko	Cyrtodactylus tamaiensis						x						DD
Tambora Bent-toed Gecko	Cyrtodactylus tambora			x									
Tanahjampea Island Bent-toed Gecko	Cyrtodactylus tanahjampea			x									
	Cyrtodactylus taungwineensis						x						
Tau't Bato Bent-toed Gecko	Cyrtodactylus tautbatorum							x					
Taybac Bent-toed Gecko	Cyrtodactylus taybacensis											x	
Tây Nguyên Bent-toed Gecko	Cyrtodactylus taynguyenensis											x	LC
Mount Tebu Bent-toed Gecko	Cyrtodactylus tebuensis					x							LC
Borikhamxay Bent-toed Gecko	Cyrtodactylus teyniei				x								DD
Thathom Bent-toed Gecko	Cyrtodactylus thathomensis				x								
Thirakhupt's Bent-toed Gecko	Cyrtodactylus thirakhupti									x			NT
Tho Chu Bent-toed Gecko	Cyrtodactylus thochuensis											x	
Thuong's Bent-toed Gecko	Cyrtodactylus thuongae											x	VU
Phnom Dalai Bent-toed Gecko	Cyrtodactylus thylacodactylus		x										
Tiger Bent-toed Gecko	Cyrtodactylus tigroides									x			LC
Banjaran Timur Bent-toed Gecko	Cyrtodactylus timur					x							LC
Tioman Island Bent-toed Gecko	Cyrtodactylus tiomanensis					x							NT
Cameron Highlands Bent-toed Gecko	Cyrtodactylus trilatofasciatus					x							EN
Moulmein Bent-toed Gecko	Cyrtodactylus variegatus						x			x			DD
Karst Forest Bent-toed Gecko	Cyrtodactylus vilaphongi				x								DD
Rakhine State Bent-toed Gecko	Cyrtodactylus wakeorum						x						DD
South Sulawesi Bent-toed Gecko	Cyrtodactylus wallacei			x									
Wangkulangkul's Bent-toed Gecko	Cyrtodactylus wangkulangkulae									x			LC
Wayakone's Bent-toed Gecko	Cyrtodactylus wayakonei				x								NT
Wel Pyan Cave Bent-toed Gecko	Cyrtodactylus welpyanensis						x						
Wetar Bent-toed Gecko	Cyrtodactylus wetariensis			x									DD
Yang Bay Bent-toed Gecko	Cyrtodactylus yangbayensis											x	LC

Common Name	Scientific name	Brunei Darussalam	Cambodia	Indonesia	Laos	Malaysia	Myanmar	The Philippines	Singapore	Thailand	Timor-Leste	Vietnam	Global Status
Yathe Pyan Cave Bent-toed Gecko	*Cyrtodactylus yathepyanensis*						x						
Sabah Lowland Bent-toed Gecko	*Cyrtodactylus yoshii*					x							LC
Ywangan Bent-toed Gecko	*Cyrtodactylus ywanganensis*						x						
Spotted Bent-toed Gecko	*Cyrtodactylus zebraicus*									x			
Chu Yang Sin Bent-toed Gecko	*Cyrtodactylus ziegleri*											x	LC
Batanta Bent-toed Gecko	*Cyrtodactylus zugi*			x									LC
Ninh Thuan Leaf-toed Gecko	*Dixonius aaronbaueri*											x	LC
	Dixonius dulayaphitakorum									x			
Orange-tailed Leaf-toed Gecko	*Dixonius hangseesom*									x			LC
Sam Roi Yot Leaf-toed Gecko	*Dixonius kaweesaki*									x			CR
	Dixonius lao				x								
Black-spotted Leaf-toed Gecko	*Dixonius melanostictus*									x			LC
Minh Le's Leaf-toed Gecko	*Dixonius minhlei*											x	LC
Cha-am Leaf-toed Gecko	*Dixonius pawangkhananti*									x			
Siamese Leaf-toed Gecko	*Dixonius siamensis*		x		x					x		x	LC
Phú Quý Island Leaf-toed Gecko	*Dixonius taoi*											x	VU
Vietnamese Leaf-toed Gecko	*Dixonius vietnamensis*		x									x	LC
Narrowhead Dtella	*Gehyra angusticaudata*									x			
Banda Island Dtella	*Gehyra barea*			x									EN
Palau Island Dtella	*Gehyra brevipalmata*			x									LC
Common Four-clawed Gecko	*Gehyra butleri*					x							DD
Fehlmann's Four-clawed Gecko	*Gehyra fehlmanni*									x		x	LC
Oudeman's Dtella	*Gehyra interstitialis*			x									DD
Kanchanaburi Four-clawed Gecko	*Gehyra lacerata*									x		x	LC
Leopold Dtella	*Gehyra leopoldi*			x									DD
Ternate Dtella	*Gehyra marginata*			x									
Common Four-clawed Gecko	*Gehyra mutilata*	x	x	x	x	x		x	x	x	x		
Pacific Dtella	*Gehyra oceanica*			x									LC
Papua Dtella	*Gehyra papuana*			x									LC
Papuan Giant Gehyra	*Gehyra serraticauda*			x									DD
Eastern Giant Stump-toed Gecko	*Gehyra vorax*			x									
Aaron Bauer's Gecko	*Gekko aaronbaueri*						x						VU
Adler's Gecko	*Gekko adleri*											x	LC
Smooth-scaled Narrow-disked Gecko	*Gekko athymus*							x					NT
Golden Gecko	*Gekko badenii*											x	EN
Boehme's Gecko	*Gekko boehmei*					x							VU
Bonkowski's Gecko	*Gekko bonkowskii*					x							VU
Brown's Forest Gecko	*Gekko browni*						x			x			LC
Cà Ná Marbled Gecko	*Gekko canaensis*											x	LC
Huu Lien Gecko	*Gekko canhi*											x	LC
Luzon Karst Gecko	*Gekko carusadensis*							x					
Chinese Narrow-disked Gecko	*Gekko chinensis*											x	LC
	Gekko cicakterbang			x		x				x			
Sibuyan Forest Gecko	*Gekko coi*							x					
Babuyan Claro Gecko	*Gekko crombota*							x					
Panay Limestone Gecko	*Gekko ernstkelleri*							x					VU

Common Name	Scientific name	Brunei Darussalam	Cambodia	Indonesia	Laos	Malaysia	Myanmar	The Philippines	Singapore	Thailand	Timor-Leste	Vietnam	Global Status
	Gekko flavimaritus									x			
Tokay Gecko	Gekko gecko	x	x	x	x	x	x	x	x	x		x	LC
Gigante Narrow-disked Gecko	Gekko gigante							x					VU
Khanh Hoa Narrow-disked Gecko	Gekko grossmanni											x	DD
	Gekko gulat							x					
Horsfield's Parachute Gecko	Gekko horsfieldii	x		x		x	x		x	x			LC
Intermediate Flying Gecko	Gekko intermedium							x					NT
Pak Djoko's Flap-legged Gecko	Gekko iskandari			x									DD
	Gekko kabkaebin									x			
Kaeng Krachan Parachute Gecko	Gekko kaengkrachanense									x			LC
Botel Gecko	Gekko kikuchii							x					DD
Kuhl's Parachute Gecko	Gekko kuhli	x		x		x	x		x	x			
Lauhachinda's Cave Gecko	Gekko lauhachindai									x			CR
Burmese Parachute Gecko	Gekko lionotum				x		x						LC
Mindoro Narrow-disked Gecko	Gekko mindorensis							x					LC
Spotted House Gecko	Gekko monarchus	x		x		x		x	x	x			
Naden Gecko	Gekko nadenensis				x								DD
Nutaphand's Gecko	Gekko nutaphandi									x			LC
Palawan Gecko	Gekko palawanensis							x					LC
Palmated Gecko	Gekko palmatus											x	LC
Sandstone Gecko	Gekko petricolus		x		x					x			LC
	Gekko popaense						x						
Batan Narrow-disked Gecko	Gekko porosus							x					LC
Reeves' Tokay Gecko	Gekko reevesii											x	
Sabah Parachute Gecko	Gekko rhacophorus					x							DD
Romblon Narrow-disked Gecko	Gekko romblon							x					LC
Ross' Calayan Gecko	Gekko rossi							x					
Russell Train's Marble Gecko	Gekko russelltraini											x	VU
Phong Nha-Ke Bang Gecko	Gekko scientiadventura											x	LC
Sengchanthavong's Gecko	Gekko sengchanthavongi				x								VU
Siamese Green-eyed Gecko	Gekko siamensis									x			LC
Smith's Green-eyed Gecko	Gekko smithii	x		x		x	x		x	x			LC
	Gekko sorok							x					DD
Ta Kou Marbled Gecko	Gekko takouensis											x	VU
Thakhek Gecko	Gekko thakhekensis				x								VU
Cambodian Flying Gecko	Gekko tokehos		x							x		x	
Three-banded Parachute Gecko	Gekko trinotaterra		x							x		x	LC
Truong's Gecko	Gekko truongi											x	DD
Vietnam Gecko	Gekko vietnamensis											x	VU
Lined Gecko	Gekko vittatus			x									
	Hemidactylus aquilonius					x	x						LC
Oriental House Gecko	Hemidactylus bowringii					x	x					x	LC
Brook's House Gecko	Hemidactylus brookii			x		x	x	x	x				LC
	Hemidactylus cf. garnotii										x		
Mocquard's House Gecko	Hemidactylus craspedotus			x		x			x	x			
Common House Gecko	Hemidactylus frenatus	x	x	x	x	x	x	x	x	x	x	x	LC

Common Name	Scientific name	Brunei Darussalam	Cambodia	Indonesia	Laos	Malaysia	Myanmar	The Philippines	Singapore	Thailand	Timor-Leste	Vietnam	Global Status
Indopacific House Gecko	Hemidactylus garnotii			x	x	x	x	x		x		x	
Burmese Spotted Gecko	Hemidactylus karenorum						x					x	LC
Murray's House Gecko	Hemidactylus murrayi					x	x						
Spotted House Gecko	Hemidactylus parvimaculatus									x			
Flat-tailed House Gecko	Hemidactylus platyurus	x	x	x	x	x	x	x	x	x	x	x	
Stejneger's Leaf-toed Gecko	Hemidactylus stejnegeri							x				x	LC
Roti Island House Gecko	Hemidactylus tenkatei			x				x			x		
Burmese Ghost Gecko	Hemidactylus thayene						x						LC
Vietnam Leaf-toed Gecko	Hemidactylus vietnamensis											x	LC
Ba Na Slender Gecko	Hemiphyllodactylus banaensis											x	LC
Spotted Slender Gecko	Hemiphyllodactylus bintik						x						LC
	Hemiphyllodactylus bonkowskii											x	
	Hemiphyllodactylus cf. typus											x	
Chiang Mai Dwarf Gecko	Hemiphyllodactylus chiangmaiensis									x			LC
Penang Island Slender Gecko	Hemiphyllodactylus cicak					x							VU
Pulau Enggano Slender Gecko	Hemiphyllodactylus engganoensis			x									
Yellow-bellied Slender Gecko	Hemiphyllodactylus flaviventris									x			
Bintang Slender Gecko	Hemiphyllodactylus harterti					x							NT
	Hemiphyllodactylus indosobrinus			x									
Philippine Slender Gecko	Hemiphyllodactylus insularis							x					DD
Khlong Lan Slender Gecko	Hemiphyllodactylus khlonglanensis									x			
Kizirian's Slender Gecko	Hemiphyllodactylus kiziriani						x						LC
	Hemiphyllodactylus kyaiktiyoensis									x			
Larut Dwarf Gecko	Hemiphyllodactylus larutensis					x							DD
Lynn-Way Slender Gecko	Hemiphyllodactylus linnwayensis											x	
Sumatran Slender Gecko	Hemiphyllodactylus margarethae			x									
Montawa Slender Gecko	Hemiphyllodactylus montawaensis									x			
	Hemiphyllodactylus nahangensis											x	
	Hemiphyllodactylus ngocsonensis											x	
	Hemiphyllodactylus ngwelwini						x						
	Hemiphyllodactylus pardalis											x	
	Hemiphyllodactylus pinlaungensis						x						
	Hemiphyllodactylus serpispecus					x							
Tebu Mountain Slender Gecko	Hemiphyllodactylus tehtarik					x							LC
Titiwangsa Slender Gecko	Hemiphyllodactylus titiwangsaensis					x							EN
Phapant Slender Gecko	Hemiphyllodactylus tonywhitteni						x						

Common Name	Scientific name	Brunei Darussalam	Cambodia	Indonesia	Laos	Malaysia	Myanmar	The Philippines	Singapore	Thailand	Timor-Leste	Vietnam	Global Status
Indopacific Slender Gecko	Hemiphyllodactylus typus	x	x			x	x	x	x	x			
Uga's Slender Gecko	Hemiphyllodactylus uga						x						
Asian Slender Gecko	Hemiphyllodactylus yunnanensis		x		x	x				x		x	LC
Ywangan Slender Gecko	Hemiphyllodactylus ywanganensis						x						
Zug's Slender Gecko	Hemiphyllodactylus zugi											x	DD
	Hemiphyllodactylus zwegabinensis						x						
Yellow-lined Smooth-scaled Gecko	Lepidodactylus aureolineatus							x					LC
Batan Scaly-toed Gecko	Lepidodactylus balioburius							x					LC
Christian Scaly-toed Gecko	Lepidodactylus christiani							x					LC
	Lepidodactylus dialeukos			x									
White-lined Smooth-scaled Gecko	Lepidodactylus herrei							x					LC
	Lepidodactylus intermedius			x									
Dark-spotted Smooth-scaled Gecko	Lepidodactylus labialis							x					LC
	Lepidodactylus lombocensis			x									
Common Smooth-scaled Gecko	Lepidodactylus lugubris	x		x	x	x	x		x				
	Lepidodactylus novaeguineae			x									LC
	Lepidodactylus oortii			x									DD
	Lepidodactylus pantai			x									
Small Broad-tailed Smooth-scaled Gecko	Lepidodactylus planicaudus							x					LC
	Lepidodactylus pollostos			x									
Sabah Scaly-toed Gecko	Lepidodactylus ranauensis					x							DD
	Luperosaurus angliit							x					
Brooks' Wolf Gecko	Luperosaurus brooksii			x									
	Luperosaurus corfieldi							x					DD
Cuming's Flapped-legged Gecko	Luperosaurus cumingii							x					DD
Jolo Flapped-legged Gecko	Luperosaurus joloensis							x					EN
	Luperosaurus kubli							x					DD
Mcgregor's Flapped-legged Gecko	Luperosaurus macgregori							x					EN
Palawan Flapped-legged Gecko	Luperosaurus palawanensis							x					DD
	Luperosaurus yasumai			x									DD
	Nactus arceo							x					
Pelagic Gecko	Nactus pelagicus			x									LC
	Nactus septentrionalis							x					
	Nactus undulatus			x									
Vankampen's Gecko	Nactus vankampeni			x									LC
Central Philippine Smooth-scaled Gecko	Pseudogekko atiorum							x					
Orange-spotted Smooth-scaled Gecko	Pseudogekko brevipes							x					VU
Zamboanga Smooth-scaled Gecko	Pseudogekko chavacano							x					
Cylindrical-bodied Smooth-scaled Gecko	Pseudogekko compresicorpus							x					
Leyte Diminutive Smooth-scaled Gecko	Pseudogekko ditoy							x					
	Pseudogekko hungkag							x					

Common Name	Scientific name	Brunei Darussalam	Cambodia	Indonesia	Laos	Malaysia	Myanmar	The Philippines	Singapore	Thailand	Timor-Leste	Vietnam	Global Status
Romblon Province Smooth-scaled Gecko	*Pseudogekko isapa*							x					
Southern Philippine False Gecko	*Pseudogekko pungkaypinit*							x					
Polillo False Gecko	*Pseudogekko smaragdinus*							x					LC
Bicol Smooth-scaled Gecko	*Pseudogekko sumiklab*							x					
Lacertidae (Eurasian Lizards)													
South-East Asian Green Grass Lizard	*Takydromus hani*											x	LC
Khasi Hills Grass Lizard	*Takydromus khasiensis*						x						
Kuhne's Grass Lizard	*Takydromus kuehnei*											x	LC
Ma Da Grass Lizard	*Takydromus madaensis*											x	DD
Asian Grass lizard	*Takydromus sexlineatus*	x	x	x	x	x				x		x	LC
Lanthanotidae (Borneo Earless Monitor)													
Borneo Earless Monitor	*Lanthanotus borneensis*	x		x		x							
Pygopodidae (Flat-footed Lizards)													
Burton's Snake Lizard	*Lialis burtonis*			x									LC
Jicar's Snake Lizard	*Lialis jicari*			x									LC
Scincidae (Skinks)													
Chinese Ateuchosaurus	*Ateuchosaurus chinensis*											x	LC
Hikida's Short-legged Skink	*Brachymeles apus*						x						LC
Bicol Short-legged Skink	*Brachymeles bicolandia*							x					
Bohol Short-legged Skink	*Brachymeles boholensis*							x					
Stub-limbed Burrowing Skink	*Brachymeles bonitae*							x					LC
Boulenger's Short-legged Skink	*Brachymeles boulengeri*							x					LC
Southern Bicol Short-legged Skink	*Brachymeles brevidactylus*							x					
Cebu Small Worm Skink	*Brachymeles cebuensis*							x					CR
Catanduañes Short-legged Skink	*Brachymeles cobos*							x					
	Brachymeles dalawangdaliri							x					
Slender Four-fingered Burrowing Skink	*Brachymeles elerae*							x					DD
Graceful Short-legged Skink	*Brachymeles gracilis*							x					LC
Hilong Short-legged Skink	*Brachymeles hilong*							x					
	Brachymeles ilocandia							x					
Aurora Short-legged Skink	*Brachymeles isangdaliri*							x					
Jessi's Slender Skink	*Brachymeles kadwa*							x					
Lapinig Islands Short-legged Skink	*Brachymeles libayani*							x					
	Brachymeles ligtas							x					
Lukban's Short-legged Skink	*Brachymeles lukbani*							x					
Robust Short-legged Skink	*Brachymeles makusog*							x					
Masbate Short-legged Skink	*Brachymeles mapalanggaon*							x					
Mindoro Short-legged Skink	*Brachymeles mindorensis*							x					
Catanduañes Limbless Skink	*Brachymeles minimus*							x					NT
Miriam's Legless Skink	*Brachymeles miriamae*									x			LC
Caraballo Mountains Loam-swimming Skink	*Brachymeles muntingkamay*							x					
Southern Burrowing Skink	*Brachymeles orientalis*							x					
Leyte Short-legged Skink	*Brachymeles paeforum*							x					
Cotabato Worm Skink	*Brachymeles pathfinderi*							x					DD
Eastern Visayas Short-legged Skink	*Brachymeles samad*							x					

Common Name	Scientific name	Brunei Darussalam	Cambodia	Indonesia	Laos	Malaysia	Myanmar	The Philippines	Singapore	Thailand	Timor-Leste	Vietnam	Global Status
Two-digit Worm Skink	Brachymeles samarensis							x					LC
Schadenberg's Burrowing Skink	Brachymeles schadenbergi							x					LC
Graceful Short-legged Skink	Brachymeles suluensis							x					
Negros Short-legged Skink	Brachymeles talinis							x					LC
Taylor's Short-legged Skink	Brachymeles taylori							x					
Western Mindanao Short-legged Skink	Brachymeles tiboliorum							x					
Negros Three-digit Worm Skink	Brachymeles tridactylus							x					LC
Tungao's Slender Skink	Brachymeles tungaoi							x					
Limbless Worm Skink	Brachymeles vermis							x					EN
Jens' Slender Skink	Brachymeles vindumi							x					
Camiguin Sur Short-legged Skink	Brachymeles vulcani							x					
Wright's Short-legged Skink	Brachymeles wrighti							x					DD
	Brachymelis sp	x											
Babar Island Rainbow-Skink	Carlia babarensis			x									
Bomberai Rainbow-Skink	Carlia bomberai			x									LC
Blue-grey-throated Rainbow-Skink	Carlia caesius			x									LC
	Carlia fusca			x									LC
White-striped Rainbow-Skink	Carlia leucotaenia			x									
Black-eared Rainbow-Skink	Carlia nigrauris			x									
Timor Rainbow-Skink	Carlia peronii			x							x		
	Carlia sp. 'Abanat River'										x		
	Carlia sp. 'Baucau'										x		
	Carlia sp. 'Maubisse'										x		
	Carlia sp. 'Meleotegi River'										x		
	Carlia sp. 'South Coast'										x		
	Carlia sp. incertae sedis										x		
Spine-eared Rainbow-Skink	Carlia spinauris			x							x		
Sukur Rainbow-Skink	Carlia sukur			x									
Halmahera Rainbow-Skink	Carlia tutela			x									
Balinese Snake-eyed Skink	Cryptoblepharus balinensis			x									
Flores Snake-eyed Skink	Cryptoblepharus burdeni			x									
	Cryptoblepharus cf. schlegelianus							x					
Lombok Snake-eyed Skink	Cryptoblepharus cursor			x									
Kei Island Snake-eyed Skink	Cryptoblepharus keiensis			x									
Leschenault Snake Eyed Skink	Cryptoblepharus leschenault			x							x		LC
New Guinea Snake-eyed Skink	Cryptoblepharus novaeguineae			x									LC
Blue-tailed Snake-eyed Skink	Cryptoblepharus renschi			x									LC
	Cryptoblepharus sp. 'Bakhita'										x		
Griffin's Keel-scaled Tree Skink	Dasia griffini							x					VU
Grey Tree Skink	Dasia grisea			x		x		x	x	x			
Olive Tree Skink	Dasia olivacea	x	x	x	x	x	x	x	x	x		x	LC
Peter's Dasia	Dasia semicincta							x	x				DD
Striped Tree Skink	Dasia vittata	x	x	x									LC
Shelford's Skink	Dasia vyneri			x		x							LC
Bronze Emo Skink	Emoia aenea			x									LC

Common Name	Scientific name	Brunei Darussalam	Cambodia	Indonesia	Laos	Malaysia	Myanmar	The Philippines	Singapore	Thailand	Timor-Leste	Vietnam	Global Status
Mangrove Skink	*Emoia atrocostata*	x		x		x		x	x			x	
Baudin's Emo Skink	*Emoia baudini*			x									LC
Bogert's Emo Skink	*Emoia bogerti*			x									LC
Brongersma's Emo Skink	*Emoia brongersmai*			x									LC
Pacific Blue-tailed Skink	*Emoia caeruleocauda*			x		x		x					LC
New Guinea Emo Skink	*Emoia callisticta*			x									LC
Copper-tailed Skink	*Emoia cyanura*			x									
Cyclops Emo Skink	*Emoia cyclops*			x									DD
Digul Emo Skink	*Emoia digul*			x									DD
Irian Emo Skink	*Emoia irianensis*			x									LC
Jamur Emo Skink	*Emoia jamur*			x									DD
Sumba Island Emo Skink	*Emoia kitcheneri*			x									
Kloss' Emo Skink	*Emoia klossi*			x									DD
Kuekenthal Emo Skink	*Emoia kuekenthali*			x									
Bourret's Emo Skink	*Emoia laobaoense*											x	DD
Shrub Whiptail-Skink	*Emoia longicauda*			x									
Great Emo Skink	*Emoia maxima*			x									LC
Boulenger's Emo Skink	*Emoia mivarti*			x									
De Vis' Emo Skink	*Emoia pallidiceps*			x									LC
Coastal Emo Skink	*Emoia paniai*			x									DD
Five-toed Emo Skink	*Emoia physicina*			x									LC
Reimschisel's Emo Skink	*Emoia reimschiisseli*			x									
Red-tailed Swamp Skink	*Emoia ruficauda*							x					DD
Dunn's Emo Skink	*Emoia similis*			x									
Sorex Emo Skink	*Emoia sorex*			x									
Four-striped Emo Skink	*Emoia tetrataenia*			x									
Antoni Night Skink	*Eremiascincus antoniorum*			x						x			
	Eremiascincus butlerorum			x									
	Eremiascincus cf. timorensis										x		
	Eremiascincus emigrans			x									
	Eremiascincus sp. 'Ermera'										x		
	Eremiascincus sp. 'Lautém'										x		
	Eremiascincus sp. 'Montane'										x		
	Eremiascincus timorensis			x									DD
Indonesian Sheen-Skink	*Eugongylus mentovarius*			x									
Bar-lipped Sheen Skink	*Eugongylus rufescens*			x									
Sula Sheen-Skink	*Eugongylus sulaensis*			x									
Luzon Montane Mabouya	*Eutropis bontocensis*							x					LC
Subic Bay Sun Skink	*Eutropis borealis*							x					
Caraga Sun Skink	*Eutropis caraga*							x					
Sapa Mabuya	*Eutropis chapaensis*											x	LC
Cuming's Mabuya	*Eutropis cumingi*							x					LC
	Eutropis cuprea							x					
Darevsky's Mabuya	*Eutropis darevskii*											x	DD
Six-striped Mabouya	*Eutropis englei*							x					DD
	Eutropis gubataas							x					

Common Name	Scientific name	Brunei Darussalam	Cambodia	Indonesia	Laos	Malaysia	Myanmar	The Philippines	Singapore	Thailand	Timor-Leste	Vietnam	Global Status
Red-lipped Sun Skink	Eutropis indeprensa			x		x	x						LC
	Eutropis islamaliit					x							
Lapu Lapu's Sun Skink	Eutropis lapulapu							x					
Lewis's Mabuya	Eutropis lewisi				x								
Long-tailed Sun Skink	Eutropis longicaudata	x		x	x					x		x	LC
Javanese Mabuya	Eutropis macrophthalma			x									
Bronze Skink	Eutropis macularia	x		x	x	x				x		x	
Keeled Skink	Eutropis multicarinata							x					
Common Sun Skink	Eutropis multifasciata	x	x	x	x	x	x	x	x	x		x	LC
Four-keelded Ground Skink	Eutropis quadricarinata						x						
Rough-scaled Brown Skink	Eutropis rudis	x		x		x	x						
Rough-scaled Sun Skink	Eutropis rugifera	x		x		x			x	x			LC
Palawan Sun Skink	Eutropis sahulinghangganan							x					
	Eutropis sibalom							x					
Black-tailed Bar-lipped Skink	Glaphyromorphus nigricaudis			x									LC
Negros Sphenomorphus	Insulasaurus arborens							x					
	Insulasaurus traanorum							x					
	Insulasaurus victoria							x					
Wright's Sphenomorphus	Insulasaurus wrighti							x					DD
Heyer's Isopachys	Isopachys anguinoides									x			LC
Lang's Limbless Skink	Isopachys borealis					x				x			LC
Gyldenstolpe's Limbless Skink	Isopachys gyldenstolpei									x			LC
Chonburi Snake Skink	Isopachys roulei									x			LC
Bipedal Skink	Jarujinia bipedalis									x			LC
White-spotted Tree Skink	Lamprolepis leucosticta			x									
Nieuwenhuis's Tree Skink	Lamprolepis nieuwenhuisii			x		x							LC
Emerald Green Tree Skink	Lamprolepis smaragdina			x				x					
	Larutia kecil					x							
Black Larut Skink	Larutia larutensis					x							VU
Single Finger Larut Skink	Larutia miodactyla					x							VU
	Larutia nubisilvicola									x			LC
Penang Island Larut Skink	Larutia penangensis					x							CR
Berumput Two-toed Skink	Larutia puehensis					x							DD
Two-lined Two-toed Skink	Larutia seribuatensis					x							NT
	Larutia sumatrensis			x									DD
Three-banded Larut Skink	Larutia trifasciata					x							NT
Osella's Skink	Leptoseps osellai									x			LC
	Leptoseps poilani											x	DD
	Lipinia auriculata									x			LC
Cheesman's Lipinia	Lipinia cheesmanae			x									DD
	Lipinia inconspicua			x									
Bornean Striped Skink	Lipinia inexpectata					x							LC
	Lipinia infralineolata			x									LC
Werner's Lipinia	Lipinia miangensis			x									DD
	Lipinia microcerca	x		x								x	
	Lipinia nitens							x					DD

Common Name	Scientific name	Brunei Darussalam	Cambodia	Indonesia	Laos	Malaysia	Myanmar	The Philippines	Singapore	Thailand	Timor-Leste	Vietnam	Global Status
Moth Skink	*Lipinia noctua*			x									
Yellow-striped Slender Tree Skink	*Lipinia pulchella*							x					LC
Four-striped Lipinia	*Lipinia quadrivittata*			x				x	x				LC
	Lipinia rabori							x					DD
Vinciguerra's Lipinia	*Lipinia relicta*			x									
Sekayu Striped Skink	*Lipinia sekayuensis*					x							LC
Semper's Tree Skink	*Lipinia semperi*							x					DD
	Lipinia septentrionalis			x									LC
Striped Lipinia	*Lipinia subvittata*			x				x					
Malaysian Striped Skink	*Lipinia surda*					x				x			DD
	Lipinia trivittata		x									x	
	Lipinia vassilievi											x	
Brongersma's Lipinia	*Lipinia venemai*			x									DD
Common Striped Skink	*Lipinia vittigera*	x	x	x	x	x		x	x			x	LC
Girard's Tree Skink	*Lipinia vulcania*							x					DD
Rusty Tree Skink	*Lipinia zamboangensis*							x					DD
Wannagong Ground-Skink	*Lobulia glacialis*			x									DD
New Guinea Four-fingered Skink	*Lygisaurus novaeguineae*			x									LC
White-spotted Supple Skink	*Lygosoma albopunctatum*						x			x			
Angel's Supple Skink	*Lygosoma angeli*		x	x						x		x	LC
Burmese Supple Skink	*Lygosoma anguinum*						x			x			LC
Bampfylde's Giant Skink	*Lygosoma bampfyldei*					x							DD
Cha Noi Supple Skink	*Lygosoma boehmei*											x	DD
Annamese Supple Skink	*Lygosoma corpulentum*				x					x		x	LC
	Lygosoma frontoparietale									x			DD
Harold Young's Supple Skink	*Lygosoma haroldyoungi*				x		x			x			LC
Even-toed Supple Skink	*Lygosoma isodactylum*	x								x			DD
	Lygosoma kinabatanganense					x							
Korat Supple Skink	*Lygosoma koratense*									x			LC
Striped Supple Skink	*Lygosoma lineolatum*						x						
Sumatra Writhing Skink	*Lygosoma opisthorhodum*			x									
	Lygosoma peninsulare					x							
Pope's Writhing Skink	*Lygosoma popae*						x						LC
Common Dotted Garden Skink	*Lygosoma punctata*											x	
Short-limbed Supple Skink	*Lygosoma quadrupes*			x									
	Lygosoma schneideri			x									
Siamese Supple Skink	*Lygosoma siamensis*		x		x	x				x		x	LC
Palawan Supple Skink	*Lygosoma tabonorum*							x					LC
Veun Sai Forest Supple Skink	*Lygosoma veunsaiensis*		x										LC
Philippine Giant Skink	*Otosaurus cumingi*							x					LC
	Papuascincus phaeodes			x									DD
	Papuascincus stanleyanus			x									
	Parvoscincus abstrusus							x					
	Parvoscincus agtorum							x					
	Parvoscincus arvindiesmosi							x					
	Parvoscincus aurorus							x					

Common Name	Scientific name	Brunei Darussalam	Cambodia	Indonesia	Laos	Malaysia	Myanmar	The Philippines	Singapore	Thailand	Timor-Leste	Vietnam	Global Status
	Parvoscincus banahaoensis							x					
Beyer's Sphenomorphus	Parvoscincus beyeri							x					NT
Boying's Zambales Mountain Skink	Parvoscincus boyingi							x					
Black-sided Sphenomorphus	Parvoscincus decipiens							x					LC
	Parvoscincus duwendorum							x					
Aurora Mountain Skink	Parvoscincus hadros							x					
Igorot Cordillera Mountains Skink	Parvoscincus igorotorum							x					
	Parvoscincus jimmymcguirei							x					
	Parvoscincus kitangladensis							x					LC
	Parvoscincus laterimaculatus							x					DD
	Parvoscincus lawtoni							x					DD
White-spotted Sphenomorphus	Parvoscincus leucospilos							x					LC
Highland Sphenomorphus	Parvoscincus luzonensis							x					NT
	Parvoscincus manananggalae							x					
	Parvoscincus palaliensis							x					
	Parvoscincus palawanensis							x					DD
	Parvoscincus sisoni							x					VU
Steere's Sphenomorphus	Parvoscincus steerei							x					LC
Aurora Mountain Skink	Parvoscincus tagapayo							x					NT
	Parvoscincus tikbalangi							x					
	Pinoyscincus abdictus							x					LC
Cox's Sphenomorphus	Pinoyscincus coxi							x					LC
Jagor's Sphenomorphus	Pinoyscincus jagori							x					LC
Leyte Sphenomorphus	Pinoyscincus llanosi							x					NT
Mindanao Sphenomorphus	Pinoyscincus mindanensis							x					NT
Chinese Skink	Plestiodon chinensis											x	LC
Shanghai Elegant Skink	Plestiodon elegans											x	LC
Four-lined Blue-tailed Skink	Plestiodon quadrilineatus	x								x		x	LC
Veun Sai Forest Supple Skink	Plestiodon tamdaoensis											x	LC
Huulien Ground Skink	Scincella apraefrontalis											x	DD
	Scincella badenensis											x	
	Scincella baraensis											x	
Darevsky's Ground Skink	Scincella darevskii											x	DD
	Scincella devorator											x	DD
Doria's Ground Skink	Scincella doriae				x					x		x	LC
Black Ground Skink	Scincella melanosticta	x	x	x						x		x	LC
Mountain Ground Skink	Scincella monticola											x	LC
	Scincella nigrofasciata			x									
Tawny Ground Skink	Scincella ochracea				x							x	LC
Burma Smooth Skink	Scincella punctatolineata						x			x			LC
Reeves's Ground Skink	Scincella reevesii		x		x	x				x		x	
Red-tailed Ground Skink	Scincella rufocaudata	x										x	
Rock-dwelling Ground Skink	Scincella rupicola					x						x	LC
	Scincella victoriana						x						LC
Double Subdigital-pads Skink	Sincella rara											x	DD
Pointed-headed Sphenomorphus	Sphenomorphus acutus							x					LC

Common Name	Scientific name	Brunei Darussalam	Cambodia	Indonesia	Laos	Malaysia	Myanmar	The Philippines	Singapore	Thailand	Timor-Leste	Vietnam	Global Status
	Sphenomorphus alfredi					x							DD
	Sphenomorphus annamiticus		x									x	
Long-toed Forest Skink	Sphenomorphus anomalopus			x		x							DD
	Sphenomorphus bacboensis											x	DD
	Sphenomorphus buenloicus											x	LC
	Sphenomorphus buettikoferi			x									DD
Cameron Highlands Forest Skink	Sphenomorphus cameronicus					x							EN
	Sphenomorphus capitolythos			x									
	Sphenomorphus celebensis			x									
	Sphenomorphus consobrinus			x									
	Sphenomorphus crassus					x							DD
Earless Forest Skink	Sphenomorphus cryptotis											x	LC
Blue-throated Forest Skink	Sphenomorphus cyanolaemus	x		x		x							LC
	Sphenomorphus dekkerae			x									
	Sphenomorphus derooyae			x									LC
	Sphenomorphus devorator											x	
Diwata Sphenomorphus	Sphenomorphus diwata							x					DD
Banded Sphenomorphus	Sphenomorphus fasciatus							x					LC
Grandison's Forest Skink	Sphenomorphus grandisonae									x			DD
Haas's Forest Skink	Sphenomorphus haasi			x		x							DD
Notaburi Forest Skink	Sphenomorphus helenae									x			DD
Brown Forest Skink	Sphenomorphus incognitus											x	LC
Indian Forest Skink	Sphenomorphus indicus		x		x	x	x			x		x	
	Sphenomorphus jobiensis			x									LC
Gunung Kinabalu Skink	Sphenomorphus kinabaluensis					x							LC
	Sphenomorphus latifasciatus					x							LC
Line-spotted Forest Skink	Sphenomorphus lineopunctulatus			x	x					x			LC
Maculated Forest Skink	Sphenomorphus maculatus		x		x	x	x			x		x	
	Sphenomorphus malaisei						x						
Malayan Forest Skink	Sphenomorphus malayanus			x									
Lesser Sunda Dark-throated Skink	Sphenomorphus melanopogon			x									LC
	Sphenomorphus meyeri			x									LC
Dwarf Forest Skink	Sphenomorphus mimicus									x		x	DD
	Sphenomorphus minutus			x									LC
	Sphenomorphus misolense			x									
	Sphenomorphus modiglianii			x									
	Sphenomorphus muelleri			x									LC
Many-scaled Forest Skink	Sphenomorphus multisquamatus			x		x							LC
Gunung Murud Forest Skink	Sphenomorphus murudensis			x		x							DD
	Sphenomorphus necopinatus			x									
	Sphenomorphus nigrolabris			x									
	Sphenomorphus oligolepis			x									
	Sphenomorphus orientalis					x							LC
	Sphenomorphus phuquocensis											x	
Blotched Forest Skink	Sphenomorphus praesignis					x				x			LC

Common Name	Scientific name	Brunei Darussalam	Cambodia	Indonesia	Laos	Malaysia	Myanmar	The Philippines	Singapore	Thailand	Timor-Leste	Vietnam	Global Status
	Sphenomorphus preylangensis		x										
	Sphenomorphus puncticentralis		x										
Sabah Forest Skink	Sphenomorphus sabanus			x		x							LC
	Sphenomorphus sananus			x									
Yellow-lined Forest Skink	Sphenomorphus sanctus			x		x							
	Sphenomorphus sarasinorum			x									
	Sphenomorphus schlegeli			x									
	Sphenomorphus schultzei			x									LC
Selangor Forest Skink	Sphenomorphus scotophilus					x				x			LC
Titiwangsa Forest Skink	Sphenomorphus senja					x							DD
Shea's Forest Skink	Sphenomorphus sheai											x	DD
Shelford's Forest Skink	Sphenomorphus shelfordi			x		x							DD
Common Forest Skink	Sphenomorphus simus					x							LC
Starry Forest Skink	Sphenomorphus stellatus		x			x				x		x	LC
	Sphenomorphus striolatus			x									
	Sphenomorphus sungaicolus					x				x			LC
	Sphenomorphus tanahtinggi			x									DD
Narrow-necked Forest Skink	Sphenomorphus tenuiculus			x		x							DD
Thai Forest Skink	Sphenomorphus tersus					x				x			LC
	Sphenomorphus tetradactylus											x	LC
Tonkin Forest Skink	Sphenomorphus tonkinensis											x	LC
Three-toed Forest Skink	Sphenomorphus tridigitus					x						x	NT
	Sphenomorphus tritaeniatus											x	NT
	Sphenomorphus tropidonotus			x									LC
Wavy-backed Forest Skink	Sphenomorphus undulatus			x									LC
	Sphenomorphus vanheurni			x									
Variable Forest Skink	Sphenomorphus variegatus			x		x		x					
	Sphenomorphus yersini											x	
	Sphenomorphus zimmeri			x									
Bowring's Supple Skink	Subdoluseps bowringii	x	x	x	x	x	x	x	x	x		x	
Herbert's Supple Skink	Subdoluseps herberti						x			x			LC
Sama Jaya Supple Skink	Subdoluseps samajaya					x							
	Subdoluseps malayana					x							
Giant Bluetongue	Tiliqua gigas			x									
Common Bluetongue	Tiliqua scincoides			x									LC
Nine-keeled Sun Skink	Toenayar novemcarinata						x	x		x			LC
Red-eyed Crocodile Skink	Tribolonotus gracilis			x									LC
Spiny Skink	Tribolonotus novaeguineae			x									LC
Bacon's Water Skink	Tropidophorus baconi											x	
Bavay's Keeled Skink	Tropidophorus baviensis											x	LC
Beccari's Water Skink	Tropidophorus beccarii	x	x	x									LC
Berdmore's Water Skink	Tropidophorus berdmorei						x			x		x	LC
Boehme's Water Skink	Tropidophorus boehmei											x	NT
Brooke's Water Skink	Tropidophorus brookei	x		x		x							LC
Indo-Chinese Forest Skink	Tropidophorus cocincinensis		x							x		x	LC
Davao Water Skink	Tropidophorus davaoensis							x					LC

Common Name	Scientific name	Brunei Darussalam	Cambodia	Indonesia	Laos	Malaysia	Myanmar	The Philippines	Singapore	Thailand	Timor-Leste	Vietnam	Global Status
Philippine Spiny Water Skink	*Tropidophorus grayi*							x					LC
Hainan Water Skink	*Tropidophorus hainanus*											x	LC
	Tropidophorus hangnam									x			DD
	Tropidophorus iniquus				x								DD
Laotian Water Skink	*Tropidophorus laotus*				x					x			LC
	Tropidophorus latiscutatus									x			DD
	Tropidophorus matsuii									x			DD
Small-scaled Water Skink	*Tropidophorus microlepis*		x	x						x			LC
Small-legged Water Skink	*Tropidophorus micropus*			x		x							DD
Misamis Water Skink	*Tropidophorus misaminius*							x					LC
	Tropidophorus mocquardii						x						
Murphy's Water Skink	*Tropidophorus murphyi*											x	LC
Nogge's Water Skink	*Tropidophorus noggei*											x	LC
Partello's Water Skink	*Tropidophorus partelloi*									x			LC
Perplexing Water Skink	*Tropidophorus perplexus*			x		x							LC
Robinson's Water Skink	*Tropidophorus robinsoni*					x				x			LC
Baleh Water Skink	*Tropidophorus sebi*					x							DD
Chinese Water Skink	*Tropidophorus sinicus*				x					x		x	LC
Thai Water Skink	*Tropidophorus thai*						x			x			LC
	Tytthoscincus aesculeticola							x					LC
Zamboanga Leaf-litter Skink	*Tytthoscincus atrigularis*							x					LC
Cursed-stone Diminutive Leaf-litter Skink	*Tytthoscincus batupanggah*					x							DD
Sulu Sphenomorphus	*Tytthoscincus biparietalis*							x					EN
Fraser's Hill Forest Skink	*Tytthoscincus bukitensis*					x							NT
Butler's Forest Skink	*Tytthoscincus butleri*					x				x			LC
Tahan Mountain Forest Skink	*Tytthoscincus cophias*					x							DD
Hallier's Forest Skink	*Tytthoscincus hallieri*	x		x		x							LC
Tioman Island Forest Skink	*Tytthoscincus ishaki*					x							LC
	Tytthoscincus jaripendek					x							
	Tytthoscincus kakikecil					x							
	Tytthoscincus keciktuek					x							
Langkawi Island Forest Skink	*Tytthoscincus lantkawiensis*					x							
Scaly-eared Diminutive Leaf-litter Skink	*Tytthoscincus leproauricularis*					x							DD
	Tytthoscincus martae					x							
	Tytthoscincus monticolus					x							
Bukit Panchor Forest Skink	*Tytthoscincus panchorensis*					x							DD
	Tytthoscincus parvus				x								
	Tytthoscincus perhentianensis					x							LC
Sibu Island Forest Skink	*Tytthoscincus sibuensis*					x							DD
Singapore Swamp Skink	*Tytthoscincus temasekensis*					x			x				
Temengor Forest Skink	*Tytthoscincus temengorensis*					x							NT
	Tytthoscincus temmincki				x								
	Tytthoscincus textus				x								
Rough Vietnamese Skink	*Vietnascincus rugosus*											x	DD

Common Name	Scientific name	Brunei Darussalam	Cambodia	Indonesia	Laos	Malaysia	Myanmar	The Philippines	Singapore	Thailand	Timor-Leste	Vietnam	Global Status
Shinisauridae (Crocodile Lizards)													
Vietnamese Crocodile Lizard	*Shinisaurus crocodilurus*											x	EN
Varanidae (Monitor Lizards)													
Peacock Monitor	*Varanus auffenbergi*			x									
Bangon Monitor	*Varanus bangonorum*							x					
Black Tree Monitor	*Varanus beccarii*			x									DD
Bengal Monitor	*Varanus bengalensis*						x						LC
Northern Sierra Madre Forest Monitor	*Varanus bitatawa*							x					
Golden-spotted Tree Monitor	*Varanus boehmei*			x									DD
Turquoise Monitor	*Varanus caerulivirens*			x									
Ceram Mangrove Monitor	*Varanus cerambonensis*			x									
	Varanus cf. salvator										x		
Kei Islands Monitor	*Varanus colei*			x									
Mindanao Water Monitor	*Varanus cumingi*							x					LC
Enteng's Monitor	*Varanus dalubhasa*							x					
Dumeril's Monitor	*Varanus dumerilii*	x	x			x	x			x	x		
Mangrove Monitor	*Varanus indicus*			x									LC
Peach-throated Monitor	*Varanus jobiensis*			x									LC
Komodo Dragon	*Varanus komodoensis*			x									VU
Biak Tree Monitor	*Varanus kordensis*			x									DD
Talaud Mangrove Monitor	*Varanus lirungensis*			x									
Panay Forest Monitor	*Varanus mabitang*							x					EN
Blue Tree Monitor	*Varanus macraei*			x									EN
Philippine Marbled Water Monitor	*Varanus marmoratus*							x					LC
Quince Monitor	*Varanus melinus*			x									
Clouded Monitor	*Varanus nebulosus*	x	x	x	x	x			x	x		x	
Western Visayas Water Monitor	*Varanus nuchalis*							x					NT
Sago Monitor	*Varanus obor*			x									
Southern Sierra Madre Forest Monitor	*Varanus olivaceus*							x					VU
Palawan Water Monitor	*Varanus palawanensis*							x					
Argus Monitor	*Varanus panoptes*			x									LC
Emerald Tree Monitor	*Varanus prasinus*			x									LC
Rainer Günther's Monitor	*Varanus rainerguentheri*			x									
Rasmussen's Water Monitor	*Varanus rasmusseni*							x					
Reisinger's Tree Monitor	*Varanus reisingeri*			x									DD
Rough-necked Monitor	*Varanus rudicollis*			x		x	x			x			
Crocodile Monitor	*Varanus salvadorii*			x									LC
Common Water Monitor	*Varanus salvator*	x	x	x	x	x	x	x	x	x		x	LC
Samar Water Monitor	*Varanus samarensis*							x					
Timor Monitor	*Varanus timorensis*			x							x		
Togian Water Monitor	*Varanus togianus*			x									
Tricolor Monitor	*Varanus yuwonoi*			x									
Zug's Monitor	*Varanus zugorum*			x									

■ ACKNOWLEDGEMENTS ■

ACKNOWLEDGEMENTS

This book could not have been possible without the help of many dedicated naturalists who worked in or visited the region. They also turned out to be talented photographers, and generously contributed photos. The list of contributing photographers can be found on p. 2. We are indebted to John Beaufoy, Rosemary Wilkinson and Krystyna Mayer for their patience and guidance throughout the project. In addition we would like to thank Parinya Pawagkhanant for reviewing an earlier draft of this book. We hope this book will inspire a cadre of naturalists who will advocate appreciating wildlife in its natural habitats and conservation of species for future generations.

FURTHER READING

Auliya, M. & Koch, A. 2020. Visual identification guide to the monitor lizard species of the world (Genus *Varanus*). Bundesamt für Naturschutz, Bonn, Germany.
Das, I. 2007. *Amphibians and Reptiles of Brunei*. Natural History Publications (Borneo).
Das, I. 2015. *Field Guide to the Reptiles of South-East Asia*. Bloomsbury (UK).
Grismer, L. L. 2011. *Lizards of Peninsular Malaysia, Singapore, and their adjacent archipelagos: their description, distribution, and natural history*. Edition Chimaira, Frankfurt am Main, Germany.
Grismer, L. L., & Quah, E. S. 2019. An updated and annotated checklist of the lizards of Peninsular Malaysia, Singapore, and their adjacent archipelagos. *Zootaxa* 4545(2), 230–248.
Grismer, L. L., Wood, P. L., Poyarkov, N. A., Le, M. D., Kraus, F., Agarwal, I. & Grismer, J. L. 2021. Phylogenetic partitioning of the third-largest vertebrate genus in the world, *Cyrtodactylus* Gray, 1827 (Reptilia; Squamata; Gekkonidae) and its relevance to taxonomy and conservation. *Vertebrate Zoology* 71, 101.
Nijman, V., Todd, M. & Shepherd, C. R. 2012. Wildlife trade as an impediment to conservation as exemplified by the trade in reptiles in Southeast Asia. Biotic evolution and environmental change in Southeast Asia, 82, 390.
O'Shea, M. 2021. *Lizards of the World: A Guide to Every Family*. Ivy Press (UK).
Sodhi, N. S., Posa, M. R. C., Lee, T. M., Bickford, D., Koh, L. P. & Brook, B. W. 2010. The state and conservation of Southeast Asian biodiversity. *Biodiversity and Conservation* 19(2), 317–328.

INDEX

Acanthosaura armata 13
  capra 14
  cardamomensis 14
  crucigera 15
  lepidogaster 16
  nataliae 16
  phuketensis 17
Aeluroscalabotes felinus 53
Agamid, Earless 17
Anglehead Lizard, Abbot's 29
  Blue-eyed 35
  Great 34
  Robinson's 43
Aphaniotis fusca 17
Banded Sphenmorph 119
Bent-toed Gecko, Agusan 63
  Annulated 63
  Balu 64
  Banded 66
  Bintang 65
  Butterfly 72
  Cryptic 66
  Doi Suthep 67
  Five-banded 70
  Four-striped 75
  Inger's 74
  Intermediate 68
  Jarujin's 69
  Lekagul's 69
  Malayan 74
  Marbled 71
  Mulu 71
  Oldham's 72
  Palawan 75
  Phetchabun 68
  Philippine 73
  Roesler's 76
  Sai Yok 76
  Sam Roi Yot 77
  Sanook 77
  Short-handed 65
  Singapore 70
  Southern Titiwangsa 64
  Spotted 80
  Tiger 78
  Tioman Island 79
  Wayakone's 80
  White-eyed 67
Blind Lizard, Flower's 52
Bronchocela burmana 18
  cristatella 18
  jubata 19

  marmorata 19
Brown Skink, Rough-scaled 112
Butterfly Lizard, Beauty 41
  Reeves' 42
  Red-banded 43
  Spotted 42
Calotes emma 20
  mystaceus 21
  versicolor 21
Cnemaspis baueri 56
  biocellata 57
  chanardi 57
  chanthaburiensis 58
  kendallii 58
  kumpoli 59
  limi 59
  mcguirei 60
  paripari 60
  peninsularis 61
  phuketensis 61
  psychedelica 62
  vandeventeri 62
Crested Dragon, Kinabalu 40
Crested Lizard, Blue 21
  Burmese Green 18
  Forest 20
  Green 18
  Marbled Green 19
Crocodile Lizard, Vietnamese 134
Crocodile Skink, Red-eyed 125
Cyrtodactylus agusanensis 63
  annulatus 63
  australotitiwangsaensis 64
  baluensis 64
  bintangtinggi 65
  brevipalmatus 65
  consobrinus 66
  cryptus 66
  doisuthep 67
  elok 67
  interdigitalis 68
  intermedius 68
  jarujini 69
  lekaguli 69
  limajalur 70
  majulah 70
  marmoratus 71
  muluensis 71
  oldhami 72
  papilionoides 72
  philippinicus 73
  pubisulcus 74

  pulchellus 74
  quadrivirgatus 75
  remidiculus 75
  roesleri 76
  saiyok 76
  samroiyot 77
  sanook 77
  tigroides 78
  tiomanensis 79
  wayakonei 80
  zebraicus 80
Dasia grisea 104
  olivacea 105
  vittata 105
Dibamus booliati 51
  floweri 52
Dixonius kaweesaki 81
  melanostictus 81
  pawangkhananti 82
  siamensis 82
Dopasia gracilis 50
Draco blanfordii 22
  cornutus 23
  cyanopterus 23
  haematopogon 24
  maculatus 24
  maximus 25
  melanopogon 25
  mindanensis 26
  palawanensis 26
  quinquefasciatus 27
  sumatranus 27
  taeniopterus 28
  volans 28
Dragon, Komodo 138
Emoia atrocostata 106
  caeruleocauda 106
  cyanura 107
  ruficauda 107
Eremiascincus antoniorum 108
Eugongylus rufescens 108
Eutropis longicaudata 109
  macularia 110
  multicarinata 111
  multifasciata 111
  rudis 112
  rugifera 112
False Garden Lizard, Bukit Larut 47
  Yellow-throated 45
False Gecko, Polillo 100
  Southern Philippine 100

■ INDEX ■

Fan-throated Lizard, Green 49
Flying Lizard, Black-bearded 25
  Blanford's 22
  Chartreuse-spotted 23
  Five-lined 27
  Horned 23
  Red-bearded 24
  Spotted 24
Forest Agamid, Flower's 46
Forest Dragon, Bell's 30
  Borneo 31
  Chameleon 32
  Peter's 33
  Philippine 36
Forest Gecko, Brown's 85
Forest Lizard, Maned 19
  Small-scaled Montane 48
Forest Monitor, Northern Sierra Madre 136
  Southern Sierra Madre 142
Forest Skink, Blotched 121
  Blue-throated 118
  Hallier's 132
  Indian 119
  Indo-Chinese 127
  Line-spotted 120
  Maculated 121
  Sabah 122
  Selangor 122
  Thai 123
  Variable 123
Four-clawed Gecko, Common 84
  Fehlmann's 83
  Kanchanaburi 83
Garden Lizard, Oriental 21
Gecko, Cat 53
  Golden 84
  Lichtenfelder's 55
  Lined 94
  Nutaphand's 90
  Palawan 90
  Palmated 91
  Phong Nha-Ke Bang 92
  Sandstone 91
  Smith's Green-eyed 93
  Tokay 86
Gehyra fehlmanni 83
  lacerata 83
  mutilata 84
Gekko badenii 84
  browni 85
  carusadensis 85

  gecko 86
  horsfieldii 87
  kaengkrachanense 87
  kuhli 88
  lionotum 88
  mindorensis 89
  monarchus 89
  nutaphandi 90
  palawanensis 90
  palmatus 91
  petricolus 91
  rhacophorus 92
  scientiadventura 92
  smithii 93
  trinotaterra 93
  vittatus 94
Glass Lizard, Burmese 50
Gliding Lizard, Barred 28
  Common 28
  Giant 25
  Mindanao 26
  Palawan 26
  Sumatran 27
Goniurosaurus araneus 54
  catbaensis 54
  lichtenfelderi 55
  luii 55
Gonocephalus abbotti 29
  bellii 30
  bornensis 31
  chamaeleontinus 32
  doriae 33
  grandis 34
  liogaster 35
  sophiae 36
Grass Lizard, Asian 101
Ground Skink, Black 117
  Doria's 117
  Reeves's 118
Harpesaurus borneensis 37
Hemidactylus brookii 94
  craspedotus 95
  frenatus 95
  garnotii 96
  platyurus 96
  tenkatei 97
Hemiphyllodactylus margarethae 97
  titiwangsaensis 98
  typus 98
Heyer's Isopachys 113
Horned Agamid, Cardamom Mountain 14

House Gecko, Brook's 94
  Common 95
  Flat-tailed 96
  Indopacific 96
  Mocquard's 95
  Roti Island 97
  Spotted 89
Hydrosaurus amboinensis 38
  pustulatus 39
Hypsicalotes kinabaluensis 40
Isopachys anguinoides 113
Karst Gecko, Luzon 85
Lamprolepis smaragdina 113
Lanthanotus borneensis 102
Leaf-litter Skink, Cursed-stone Diminutive 132
Leaf-toed Gecko, Black-spotted 81
  Cha-am 82
  Sam Roi Yot 81
  Siamese 82
Leiolepis belliana 41
  guttata 42
  reevesii 42
  rubritaeniata 43
Leopard Gecko, Cat Ba 54
  Chinese 55
  Vietnamese 54
Lepidodactylus lugubris 99
  ranauensis 99
Lialis jicari 103
Lipinia vittigera 114
Long-headed Lizard, Khao Nan 47
Lygosoma angeli 114
  corpulentum 115
  koratense 115
  quadrupes 116
Malayodracon robinsonii 43
Mantheyus phuwuanensis 44
Monitor, Bengal 135
  Borneo Earless 102
  Clouded 141
  Dumeril's 137
  Quince 140
  Rough-necked 144
  Timor 146
Mountain Agamid, Burmese 46
Narrow-disked Gecko, Mondoro 89
Night Skink, Antoni 108
Nose-horned Lizard, Bornean 37
Otosaurus cumingii 116
Parachute Gecko, Burmese 88

INDEX

Horsfield's 87
Kaeng Krachan 87
Kuhl's 88
Sabah 92
Three-banded 93
Pelturagonia nigrilabis 44
Phu Wua Lizard 44
Physignathus cocincinus 45
Pricklenape, Boulenger's 15
  Brown 16
  Green 14
Pseudocalotes flavigula 45
  floweri 46
  kakhienensis 46
  khaonanensis 47
  larutensis 47
  microlepis 48
Pseudogekko pungkaypinit 100
  smaragdinus 100
Ptyctolaemus gularis 49
Rock Gecko, Bauer's 56
  Chan-ard's 57
  Chanthaburi 58
  Fairy 60
  Kendall's 58
  Kumpol's 59
  McGuire's 60
  Peninsular 61
  Phuket 61
  Psychadelic 62
  Tioman Island 59
  Twin-spotted 57
  Van Deventer's 62
Sailfin Lizard, Amboina 38
  Philippine 39
Scaly-toed Gecko, Sabah 99
Scincella doriae 117
  melanosticta 117
  reevesii 118
Sheen Skink, Bar-lipped 108
Shinisaurus crocodilurus
  vietnamensis 134
Shrub Lizard, Black-lipped 44
Skink, Bronze 110
  Copper-tailed 107
  Gunung Kinabalu 120
  Keeled 111
  Mangrove 106
  Pacific Blue-tailed 106
  Philippine Giant 116
Slender Gecko, Indopacific 98
  Sumatran 97

Titiwangsa 98
Smooth-scaled Gecko, Common 99
Snake Lizard, Jicar's 103
Sphenomorphus cyanolaemus 118
  fasciatus 119
  indicus 119
  kinabaluensis 120
  lineopunctulatus 120
  maculatus 121
  praesignis 121
  sabanus 122
  scotophilus 122
  tersus 123
  variegatus 123
Spiny Lizard, Natalia's 16
Striped Skink, Common 114
Subdoluseps bowringii 124
  herberti 124
  samajaya 125
Sun Skink, Common 111
  Long-tailed 109
  Rough-scaled 112
  Supple Skink, Angel's 114
  Annamese 115
  Bowring's 124
  Herbert's 124
  Korat 115
  Sama Jaya 125
  Short-limbed 116
Swamp Skink, Red-tailed 107
Singapore 133
Takydromus sexlineatus 101
Tree Agamid, Phuket Horned 17
Tree Lizard, Peninsular Horned 13
Tree Monitor, Blue 139
  Emerald 143
Tree Skink, Emerald Green 113
  Grey 104
  Olive 105
  Striped 105
Tribolonotus gracilis 125
Tropidophorus beccarii 126
  brookei 126
  cocincinensis 127
  grayi 127
  laotus 128
  microlepis 128
  micropus 129
  misaminius 129
  partelloi 130
  robinsoni 130

sebi 131
sinicus 131
Tytthoscincus batupanggah 132
  hallieri 132
  temasekensis 133
Varanus bengalensis 135
  bitatawa 136
  cumingi 137
  dumerilii 137
  komodoensis 138
  macraei 139
  marmoratus 140
  melinus 140
  nebulosus 141
  nuchalis 141
  olivaceus 142
  palawanensis 142
  prasinus 143
  rudicollis 144
  salvator 145
  timorensis 146
Water Dragon, Chinese 45
Water Monitor, Common 145
  Mindanao 137
  Palawan 142
  Philippine Marbled 140
  Western Visayas 141
Water Skink, Baleh 131
  Beccari's 126
  Brooke's 126
  Chinese 131
  Laotian 128
  Misamis 129
  Partello's 130
  Philippine Spiny 127
  Robinson's 130
  Small-legged 129
  Small-scaled 128
Worm Lizard, Boo Liat's 51

Other books about the wildlife of Southeast Asia from John Beaufoy Publishing

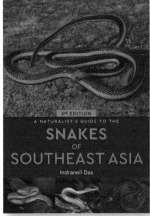

See our full range at www.johnbeaufoy.com